THE COMPLETE BOOK OF

Jaguar

THE COMPLETE BOOK OF

Jaguar

Every Model Since 1935

NIGEL THORLEY

motorbooks

Quarto.com

© 2019 Quarto Publishing Group USA Inc.
Text © 2019 Nigel Thorley

First Published in 2019 by Motorbooks, an imprint of The Quarto Group,
100 Cummings Center, Suite 265-D, Beverly, MA 01915, USA.
T (978) 282-9590 F (978) 283-2742 QuartoKnows.com

Motorbooks titles are also available at discount for retail, wholesale, promotional, and bulk purchase. For
details, contact the Special Sales Manager by email at specialsales@quarto.com or by mail at The Quarto
Group, Attn: Special Sales Manager, 100 Cummings Center, Suite 265-D, Beverly, MA 01915, USA.

23   6

ISBN: 978-0-7603-6390-4

Digital edition published in 2019

Library of Congress Cataloging-in-Publication Data available

Acquisitions Editor: Zack Miller
Art Director: Laura Drew
Design and page layout: Laura Shaw Design

Printed in China

CONTENTS

PREFACE

There's always been something special about the Jaguar brand, even at the very start before World War II when the name was just getting established. At the time, it stood for sportiness and excellent value, attributes that continue to this day.

In the postwar years, Jaguar's slogan, *Grace . . . Space . . . and Pace . . .*, said it all. The incredible success the brand enjoyed from those early years escalated the cars' esteem and reputation to become a world leader in quality, performance, and value. Success in world sports car racing furthered the brand's acclaim, with no fewer than seven outright wins at the Le Mans 24-hour endurance race, and many other feats in racing and competitive rallying.

Those achievements, particularly in the 1950s, aided by an expanding range of superb new models, led to company founder Sir William Lyons receiving a Knighthood for his work in the British motor industry. Success led to further success with the introduction of an exciting range of new cars like the E-Type sports and Mark 2 compact saloon in the 1960s, and then into the XJ era, a model continued to this day.

Thanks to the ups and downs of the automotive industry and world trade over the years, Jaguar has been through the mill a few times. Merging with the British Motor Corporation (BMC) in 1968 to form British Motor Holdings, then going through the traumas of public ownership and British Leyland, and eventually into privatization and then Ford ownership in the 1990s.

The business (along with Land Rover) is now in the hands of the Indian Tata Group, which continues to develop the Jaguar reputation for "Best of British" engineering and styling, one of the few British manufacturers to have sustained all that has been thrown at it over the years. Success has bred success.

This book celebrates Jaguar's ability to create stylish and exciting cars from its early beginnings, right up to the present day.

WHAT'S IN A NAME?

In British motoring terms, the 1920s was a period of motorcycle domination. Although the car industry was growing quickly, only the well-heeled could afford the extravagance of an expensive car. Even most of the middle classes had to settle for basic models if they wanted four wheels. To enterprising eyes, there was an intermediate option left unfilled.

In 1922, William Lyons was a young man selling motorcycles in his hometown of Blackpool. William Walmsley, who lived on Lyons's street, had a project suited to Lyons's trade underway in his family's rear garden. Walmsley's Zeppelin-shaped sidecar was a stylish means to give a motorcycle additional passenger or cargo room.

Lyons initially asked Walmsley to build him one. Thinking further, Lyons proposed that the two Williams go into business together manufacturing sidecars for the motoring public. Walmsley agreed to give it a go. With a little financing from both sets of parents, they set up the Swallow Sidecar Company for £200 in a rented space in Lancashire.

Walmsley ran the manufacturing side "hands on" and Lyons handled sales, marketing, and administration. The business blossomed. They had their first display stand at the London Motorcycle Show in 1924 and eventually had to move to larger premises (also in Blackpool) as they added several new designs to their range.

In 1926, they expanded the name of the business to the Swallow Sidecar & Coachbuilding Company, highlighting a change in direction. Carrying out repairs on a local owner's Austin Seven, Lyons realized the potential to build stylish bodies on this type of chassis. The pair purchased an Austin Seven chassis and with the help of their factory craftsmen,

they built a prototype Austin Seven Swallow based on Lyons's styling.

Lyons showed the car to one of the UK's most prominent automobile distributors. They were so impressed they offered to handle the supply of chassis directly from the Austin Motor Company and to sell the finished cars on behalf of Swallow. Within a year, production was underway and the little Austin Seven Swallow became a success, both in its original coupe form and as a saloon version.

Manufacturing could not keep up with the sales. Situated on the west coast of England, Swallow didn't have ready access to the components needed for the cars (much of which were manufactured

Taken from the original SS brochure for 1931, the SS1 combined Standard Motor Company mechanics with a William Lyons–styled body to create a £310 marvel that looked much more prestigious and faster than it really was.

in the West Midlands, miles away), nor to a sufficiently qualified workforce to build them. In 1928, Lyons took the decision to move the whole business (including most of the employees) to the Midlands, the heart of the British motor industry.

The move led to an expansion of production (and sales) and the bodying of other cars in the same style, including Morris, Wolseley, and particularly Standard, then one of the area's big five producers.

Standard Motor Company began supplying Swallow with modified versions of their Standard Nine and Sixteen chassis to be bodied by Swallow and sold under a new brand name: SS (often misattributed as standing for "Standard

Swallow," the name was actually an abbreviation for "Swallow Sports"). This new name in the automobile industry was launched at the 1931 British Motor Show as the SS1 (six-cylinder) and SS2 (four-cylinder) two-door coupes serving as a beginning of Lyons's strategy to expand more broadly into the motorcar business.

The two new models were very well received, leading to an expansion and development of the range from 1933 with longer chassis, more powerful engines, and different body styles, but all based on the same basic mechanical structure.

Although the sidecar part of Swallow's business was still thriving, the success of the SS cars led to the forma-

tion of a separate company in 1934: SS Cars Limited. At the same time, differences between Walmsley and Lyons resulted in the former being bought out of the business and Lyons taking full control of both the automotive and sidecar businesses.

Lyons wanted to produce more of the car himself, with new models to emulate the success and reputation of established British brands like Alvis, Lagonda, and even Bentley. This new style of car got a new name in 1935, which would eventually replace SS: Jaguar.

CHAPTER ONE

MODEL TO MARQUE

The Early Pushrods

Spurred by the success of his Standard-based SS models, William Lyons set about designing an entirely new car that would require a new name, although at this point not a brand name in its own right, but merely a model name epitomizing the essence of the car. Used in contemporary culture, and in pre-Columbian Central and South America, as a symbol of power and strength, the jaguar's speed and predatory nature made it a widely feared animal. The stealthy and streamlined look of the pouncing cat was the perfect emblem for Lyons's new car.

1935: Introducing a New Car and a New Name.

Lyons's Jaguar was a completely new design and a very advanced project for such a fledgling company. The first aspect Lyons focused on was the chassis. Produced for SS by the Standard Motor Company, but to Lyons's own design, the chassis would accept a new four-door sports saloon body style, also designed by him.

Standard's engines produced insufficient power for Lyons's needs, so he employed the services of Harry Weslake,

one of the foremost motoring engineers of the day. Weslake redesigned the Standard side-valve engines to an overhead valve arrangement, improving their efficiency and performance considerably. Standard agreed to manufacture two new engines—a 2.5-liter six cylinder and a 1.5-liter four cylinder—on the understanding that the six would be exclusively for SS use, but that the smaller unit could be used in forthcoming new Standard models as well.

Offering an all-new chassis, stylish sports saloon bodywork, and the choice

of the two engines, the new SS Jaguar models were launched to an audience of respected British motor industry experts and journalists at the Mayfair Hotel in London in September 1935. After viewing the 2.5-liter model on a plinth in the dining room and being told of its specification, the tables were supplied with paper and pencils for the gathering to estimate the price at which this new car was to be sold in 1936. No one in attendance came close to the true figure of £385. In fact, few believed it possible to produce such an advanced car at that

(ABOVE) The first-generation SS Jaguar 2.5-liter saloon as it was launched from 1936, with a side-mounted spare wheel that would have had a cover originally. The radiator mascot was an option.

(LEFT) The overhead-valve six-cylinder engine designed by Harry Weslake for SS, seen here in 2.5-liter form.

price. Lyons proved them wrong and the SS Jaguar was a success.

With the introduction of the SS Jaguars, production of all SS1 and SS2 cars ceased except for the two-door, six-cylinder Tourer, which was renamed SS Jaguar Tourer and fitted with the overhead-valve 2.5-liter engine.

1936–1937: The New Kid on the Block

Production of the new models got underway immediately, initially with the two four-door saloon models and then with the addition of a two-seater sports car (derived from the mechanics of the saloons): the SS Jaguar 100. This rakish sports car was very much styled on similar sports cars of the period (including a low-production SS90 based on the SS1), with its swept-back wing design, prominent chromed radiator grille surround, and tripod headlights. It gave an incredible visual impact. Initially there was only one 100 model with the 2.5-liter engine.

Although produced in small numbers, the SS Jaguar 100 created considerable

buzz for the company and its products, enhancing its reputation for fine cars.

1937–1940: Metal to Metal

Initial sales of the saloons were promising, so much so that there was a need to streamline production. Demand would increase further with the provision of extra models. The renamed Tourer was discontinued after only 125 were produced, as it didn't fit with production changes the company was about to implement.

The SS Jaguar saloons were initially built in the traditional expensive and time-consuming manner of steel panels grafted onto an ash wooden framework. All-steel bodyshells were the way forward, which SS adopted late in 1937.

At the same time, SS introduced two-door drophead coupe versions of all the saloons, but as the numbers produced would be much smaller and a degree of handcrafting was required, traditional build methods continued for these cars.

SS also introduced a third new engine, a 3.5-liter, to the range, based on the

six-cylinder unit. This engine was very much a high-performance model capable of propelling cars to speeds above 90 miles per hour. Available for the 1938 model year, the 3.5-liter found a home in the saloon and drophead coupe, as well as the SS Jaguar 100 sports.

The final upgrade to the model line came when displacement of the 1.5-liter four-cylinder engine was increased to 1.75 liters.

All models continued in production alongside sidecar manufacture until the outbreak of World War II, when civilian vehicle production was turned over to the war effort.

The company's first era ended with its move from Lancashire to Warwickshire. Since the introduction of the model name Jaguar in 1935, the company had produced a total of 14,359 cars. Numerous changes awaited after World War II, not least the adoption of Jaguar as the company and brand name.

The new chassis designed to accept the overhead-valve engine and the stylish saloon coachwork.

(ABOVE) The beautiful SS Jaguar 100 sports car was available for a limited period prewar with either the 2.5- or 3.5-liter engine.

(LEFT) By the end of 1937, a larger-engined 3.5-liter model was added to the range, requiring a longer bonnet. At the same time, all models lost the side-mounted spare wheel.

(**LEFT**) The early instrument layout for prewar models. Note, too, the arrival of the four-spoke steering wheel that became a traditional feature on the company's cars for many years.

(**BELOW**) The 1.5-liter (later 1.75-liter) SS Jaguar with a shallower and lower bonnet that makes the side-mounted spare wheel stand proud. The metal spare-wheel cover was standard on all these cars until the introduction of the steel-bodied versions.

(**OPPOSITE**) One of the virtually hand-built drophead coupe models from the SS Jaguar range, initially available in all three engine sizes. Here the roof is unfurled into the sedanca position, allowing the driver and front seat passenger to take in the fresh air while leaving the rear seat passengers with cover.

SPECIFICATIONS

MODEL	SS Jaguar 1.5-liter Saloon	SS Jaguar 2.5-liter Saloon	SS Jaguar 3.5-liter Saloon	SS100 2.5-liter Saloon	SS100 3.5-liter Saloon
ENGINE SIZE	1,608cc	2,663cc	3,485cc	2,663cc	3,485cc
CARBURETION	Single Solex	2 x 1.25 SU	2 x 1.5 SU	2 x 1.5 SU	2 x 1.5 SU
MAX BHP	52@4,300	102@4,600	125@4,500	102@4,600	125@4,250
MAX TORQUE	Not stated	Not stated	Not stated	Not stated	Not stated
GEARBOX	4-speed	4-speed	4-speed	4-speed	4-speed
AUTOMATIC	n/a	n/a	n/a	n/a	n/a
0 TO 60 MPH	33 sec.	17.5 sec.	10 sec.	12.8 sec.	10.9 sec.
STANDING ¼ MILE	24 sec.	20 sec.	19.4 sec.	18.6 sec.	17.1 sec.
TOP SPEED	70 mph	85 mph	92 mph	94 mph	101 mph
AVERAGE FUEL CONSUMPTION	25 mpg	21 mpg	18 mpg	20 mpg	20 mpg

CHAPTER TWO

POSTWAR PUSHROD YEARS

Big changes arrived on the British automotive scene after World War II. Manufacturers that had come through the hostilities had to adapt to a different world, where it was vital for a virtually bankrupt Great Britain to export and bring in needed funds.

SS cars and the Swallow Sidecar & Coachbuilding Company had done well with their contributions to the war effort. Postwar, it was time to concentrate on reinstating car production, developing new models, and, most significantly, regrouping and rebadging. The SS insignia now meant something very different to the world and was no longer viable. It was time to install Jaguar as the brand name and sell off the sidecar business—Jaguar Cars Ltd. was born.

1945–1947: The Same, Only Similar

Production resumed immediately with the reintroduction, initially, of the prewar 1.75-liter saloon, of which 141 were assembled. Into 1946 the range was expanded with the reintroduction

of the two prewar six-cylinder saloons, for a total production that year of 2,928 cars. It wasn't until the end of 1947 that drophead coupes were reintroduced, but only in six-cylinder form.

With resources lean and demand high, Jaguar made only minor changes to exterior trim in the reintroduced prewar models. The waistline chrome trims were thinner and the radiator shells were subtly different, but not interchangeable with the prewar design. The scuttle panel, immediately in front of the windscreen, was made a couple of inches wider, which reduced the length of the bonnet. This allowed for a larger air intake flap and made space for a more modern radio underneath the instrument panel. Of course, the badging was altered, too. Off came the SS insignia, replaced by the word "Jaguar."

Mechanically, the six-cylinder cars were equipped with a hypoid rear axle (the four-cylinder car had always featured this). With a change of ratios this improved performance a little. Minor changes were also made to the engine and the 1.75-liter four-cylinder received a new water-heated inlet manifold to improve breathing and brake horsepower.

The braking system for all models came in for revision with the fitment of a Girling two-leading-shoe design, although still rod-operated (a system that was already antiquated).

There were also changes to the interior. The occasional "picnic" tables fitted to the rear of the front seats were eliminated. Also, the plain leather upholstered seats were replaced with a pleated style. There was also a new style dashboard layout.

(ABOVE) Postwar production got underway with effectively the same models as prewar, though the badging had changed and exterior trim was simpler.

(LEFT) The prewar rear bumper carried the SS logo on a center plinth. From 1946, it was a simple J.

A new interior layout for the dashboard and a return of pleats to the seating.

These postwar Jaguars later became unofficially known as Mark IVs. This is the 3.5-liter model with the longer bonnet.

The excellent SS Jaguar 100 never returned to production. It was not considered viable in light of resources needed to expand saloon production to meet demand.

In this immediate postwar period, amid myriad detail changes, the biggest single change, other than the adoption of Jaguar as the brand name, was in pricing. The cost of rare materials and inflation hit the UK motor industry hard. For example, an SS Jaguar saloon that would have cost £400 before the war, now cost nearly £1,000 (if you could even find one in the UK!). Priority was given to overseas sales, so the British home market didn't see many of the early postwar models.

Nevertheless, these postwar cars did much to establish the Jaguar name at home and abroad, helping the company enter lucrative markets overseas. Much more was yet to come.

1948: A New Model Is Born

By the end of 1948, Jaguar Cars Ltd. had produced another 11,658 of its prewar-designed models, plus a further

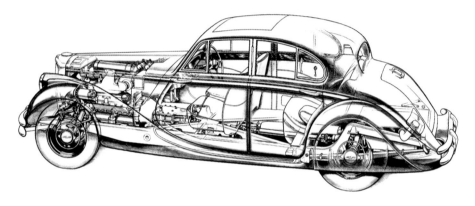

The 1948 Mark V was more new than old, but still very traditional. Jaguar's new chassis was mated to a facelifted prewar-styled body with smoother lines and a new interior.

311 examples in 1949, closing out the development of those cars. However successful these cars had been (total production of SS Jaguar and postwar Jaguar models was 26,017 at this point), William Lyons was not satisfied. He still wanted to build a modern, stylish sports saloon capable of 100 miles per hour. That would mean a *new* car in more than just name.

Work progressed on all aspects of that new saloon, but certain elements were not complete. A new engine designed and built by Jaguar specifically for the car was just about ready for

production, but another vital element— the exciting new body—was not.

Independent front suspension (IFS) was still a rarity on production cars at this time, although the French manufacturer Citroën had devised a successful system for their Traction Avant range in the 1930s. Jaguar had experimented with IFS systems prewar and now felt it was a vital move forward for the new chassis. Made of top and bottom wishbones carrying ball-jointed stub axles, with springing by longitudinal torsion bars with telescopic shock absorbers, the system featured rubber bushings in all pivots to cut down vibration. The rubber-mounted antiroll bar was secured under the front crossmember running between the two lower suspension beams. The IFS was advanced and successful—so much so that it continued virtually unchanged for all Jaguar models through the 1950s and for many models into the 1960s, a tribute to the prowess of Jaguar engineers.

For the rear suspension, Jaguar used softer and longer springs along with Girling lever-arm shock absorbers mounted on the side members ahead of the rear axle. For this new chassis, 16-inch-diameter wheels with wider rims replaced the previous 18-inch wheels. This lowered the center of gravity, improving balance of handling. A Burman recirculating ball system was retained for the steering.

The braking was updated to a hydraulic system with 12-inch drums of Girling two-leading-shoe type. The handbrake operated normally on the rear wheels but now from an umbrella-style handle mounted below the dashboard near the driver.

The new chassis was very strong and rigid with large box sections, crossmembers, and a pronounced lift over the rear-axle area. As this new chassis was ready for production, Lyons decided that it would form the basis of an interim model to maintain sales momentum for the next couple of years until the entirely new saloon was ready.

The chassis was of course designed to take the new engine. However, as Jaguar had purchased the tooling from Standard to build their own pushrod engines after the war, and to ensure the new saloon, when it eventually came out, would have the most impact, Jaguar installed the six-cylinder pushrod units in this interim model (Standard had retained the four-cylinder tooling for their own use). The 2.5- and 3.5-liter engines were carried over virtually unchanged, and with

The chassis designed for an entirely new saloon in the 1950s was launched with the pushrod engine in the Mark V from 1948.

Slowly, Jaguar's dashboard arrangements progressed with yet more changes from the Mark IV. The new version of the Bluemels four-spoke steering wheel was also featured on the XK120 sports.

the same gearbox, although a divided propshaft was used for the first time.

The old-style bodywork received a major workover. The bonnet sloped, the profile of the radiator grille differed, the headlights were flared into the front wings, more substantial bumper bars were fitted, spats covered the rear wheels, wider doors had separate chrome-on-brass window frames, and the screen pillars were slimmer with a taller windscreen. Push-button door locks were used for the first time on a British car.

Internally, it was still very traditional, with plain (unpleated) leather trim, adjustable-height front bucket seats, and a conventional bench seat for the rear. Walnut veneer was used extensively. The businesslike dashboard had a complete complement of instruments, and the cars retained a large Bluemels steering wheel with its well-known four-spoke design.

The chromed window surrounds incorporated larger front quarterlights and a complex bellcrank system for the rear quarterlights, a feature that would

(**ABOVE**) The Mark V as depicted in the sales brochure. Flared-in front lighting, elegant sweeping lines with more chromed adornment, and an incredibly successful transformation from the prewar design.

(**RIGHT**) Produced in small numbers, the drophead version of the Mark V was arguably the most attractive of all the models the company had produced up to that time.

become a recognized design on all Jaguar saloons for many years. The interior door handles were a new, larger trigger type, and the doors retained their prewar center-hinged style.

This "new" model was named the Mark V because it was the fifth prototype in the development of the independent front suspension and Jaguar needed a new identity to avoid confusion with the prewar designs.

The Mark V saloon and 3.5-liter drophead coupe were launched in October 1948. Arguably the more stylish of the Mark V models, the latter was built in very small numbers, as it was still largely handmade.

1949–1951: Mark V Production Underway

It wasn't until 1949 that postwar production got properly underway with a total of 3,763 saloons and 9 dropheads turned out.

Few changes took place during the production period apart from in-service upgrades to rubber sealing, and the relocation of the front and rear ashtrays.

The engines initially were fitted with alloy connecting rods, but these were changed to steel during production, and the braking system came in for some amendments.

Mark V sales peaked in 1950, with a total of 5,679 produced, followed in 1951 by 1,030. A 2.5-liter-engined version of the drophead coupe became available just for 1950, but only 29 were made.

For an interim model, the Mark V offered a unique style, brought the prewar designs up to date, and was well-accepted, particularly in overseas markets. The first Jaguar sold with the slogan: *Grace . . . Space . . . and Pace . . .* , the Mark V measured up.

Bucket-like front seating in the Mark V saw a return to unpleated leather upholstery.

The rear compartment of the Mark V was the most luxurious yet from Jaguar. Note the then-new trigger door locks on the door panels. The bellcrank-styled quarterlight catches are also visible, yet another feature to follow through on so many other models.

SPECIFICATIONS

MODEL	Jaguar 1.75-liter Saloon	Jaguar 1.75-liter Saloon	Jaguar 3.5-liter Saloon	Mark V 2.5-liter Saloon	Mark V 3.5-liter Saloon
ENGINE SIZE	1,776cc	2,633cc	3,485cc	2,663cc	3,485cc
CARBURETION	1 x SU	2 x 1.25 SU	2 x 1.5 SU	2 x 1.5 SU	2 x 1.5 SU
MAXIMUM BHP	65@4,600	105@4,600	125@4,500	105@4,600	125@4,250
MAXIMUM TORQUE	Not stated	Not stated	Not stated	Not stated	Not stated
GEARBOX	4-speed	4-speed	4-speed	4-speed	4-speed
AUTOMATIC	n/a	n/a	n/a	n/a	n/a
0 TO 60 MPH	25 sec.	17 sec.	10 sec.	18 sec.	14.7
STANDING ¼ MILE	n/a	20 sec.	19.4 sec.	20 sec.	20 sec.
TOP SPEED	72 mph	87 mph	92 mph	87 mph	90 mph
AVERAGE FUEL CONSUMPTION	27 mpg	19 mpg	18 mpg	19 mpg	18 mpg

CHAPTER THREE

THE STRAIGHT-SIX XKs

The British motor industry changed quickly after World War II to entice world markets with new and exciting cars. Key to this evolution was an uprated engine capable of serving a new range of cars for a period of many years. Designated the XK, that engine was originally intended for a luxury saloon that wouldn't go into production until 1950. The engine had such promise, however, that Jaguar chose to showcase the XK in a stylish low-production sports car in 1948 at the first British Motor Show in London.

1948: The Launch Pad

The XK120 was one of Britain's most significant sports cars. Its ongoing development and racing successes at Le Mans in the 1950s elevated not only Jaguar but the whole British car industry.

With work advanced on the XK engine and the Mark V saloon forthcoming, William Lyons decided that a low-production sports car utilizing the new engine could generate some marketing buzz and test the waters for the Mark VII saloon, the car destined to have the new engine.

Assembly of the concept for the 1948 British Motor Show took place only months beforehand. The new chassis (for the Mark V/VII saloon) was cut down by 1½ feet at the center section; the cross brace was replaced with a single box section crossmember. The overall frame was also narrowed slightly. The suspension, steering, and braking systems were the same as the Mark V.

The new 3,442cc XK six-cylinder engine was fitted to the chassis with the twin SU carburetors and matched to a

Moss four-speed gearbox via a one-piece propshaft to the Mark V rear axle.

The distinctive bodyshell, created by William Lyons, had more than a hint of prewar BMW 328 Mille Miglia special to the design. It produced quite a stir at the show, with an all-enveloping, stylish, curvaceous body, swept-back front wings, and fully enclosed rear wheels.

The body was built entirely of handcrafted aluminum, elegant by any standards, and even aerodynamic. Available only as a roadster at this point, it had a one-piece "alligator style"

(ABOVE) The original XK120 roadster's pure lines captivated the public on its debut in 1948. It was a car without equal in terms of performance and style.

(LEFT) The legendary Jaguar XK six-cylinder twin-camshaft engine launched in the XK120 in 1948 and fitted in some vehicles up until 1992. Total production reached over 800,000 units.

THE "XCEPTIONAL" XK

Typical installation of the XK engine, a unit made to look as good as it would perform. Access under the alligator bonnet was somewhat hampered by the taper of the wings and the attached grille.

SS had produced several prototype engines, prewar, with twin overhead camshafts and hemispherical combustion chambers. The last of these prototypes, coded XJ, was the forerunner of the six-cylinder XK engines.

For lightness, engineers conceived an aluminum cylinder head, with valves 70 degrees to vertical, operated by twin chain-driven overhead camshafts. The cast-iron block carried a very strong seven-bearing crankshaft with large main bearings and a gear-driven oil pump.

The fuel-air mixture was supplied by two 1.75-inch SU carburetors on an aluminum inlet manifold. Exhaust exited two three-branch vitreous-enameled manifolds joining into a single downpipe and tailpipe.

The six-cylinder engine displaced 3,442cc and was rated at 160 brake horsepower at 5,400 rpm. For looks to match its performance, the engine boasted a polished aluminum intake manifold and camshaft covers and enameled exhaust manifolds.

The outstanding performance inherent in the XK six-cylinder engine is reflected in its exceptional longevity and versatility. The design was reconfigured to other capacities (2.4, 2.8, 3.8, and 4.2 liters), fuel-injected, and adapted for commercial, military, and marine use. It powered every Jaguar car from the XK120 sports and Mark VII saloon, through all 1950s and 1960s models, continuing in production through 1992. More than half a million were produced.

bonnet incorporating an oval grille. The doors were sculptured to allow for better elbow room, and conventional side screens could be mounted on the doors. It had a split windscreen, while a one-piece removable roof provided weather protection and could be stored behind the seats.

Internally, the car was well finished for a two-seater, with more space than usual in a sports car. It featured wide, pleated-leather seats, quality carpeting, and leather-clad door panels with map pockets. Jaguar retained the now-familiar Bluemels four-spoke steering wheel and workmanlike dashboard with a full complement of instruments and controls. A short, stubby gear lever was centrally floor mounted, and the handbrake was of the fly-off type, a racey feature of many sports cars.

The civilized accommodations continued with a good luggage area; under the boot floor, the spare wheel and tools were easy to access.

The car was named the XK120 Super Sports (XK after the engine and 120 the suggested top speed). The big reveal came at the British Motor Show on October 27 to world acclaim. It was priced at £998 plus British purchase tax—the same price as Jaguar's other new model, the Mark V.

Jaguar had intended to launch a four-cylinder variant of the car, dubbed XK100, at the same price. Given the interest in the XK120 and the lack of price differential, however, no completed examples of the XK100 were produced and the idea was dropped.

Small production numbers for the XK120 had to be abandoned after the first day of the motor show as interest was so great. It was an incredible start to the model, a car whose body had reputably been conceived in a matter of weeks prior to the show. Its success was guaranteed, and Jaguar had to gear up for serious production.

(LEFT) Duotone leather trim was the norm for the XK120. The dashboard layout is simple but meets all the instrument needs of a spirited driver. The four-spoke Bluemels steering wheel became a trade-mark feature of all XKs (and many other Jaguars up until the 1960s). This car is equipped with the white steering wheel, an extra-cost option fitted to many cars destined for the US.

(RIGHT) The XK120 takes pride of place on the Jaguar stand at the 1948 Earls Court Motor Show in London, alongside the new Mark V saloon. The crowd says it all!

1949: Production Underway

Production of the partly hand-finished XK120s began late in 1949 on the same line as the Mark V saloons. A mere 97 examples were produced that year, with most production shipping overseas.

Because the model name was based on an estimation of the car's top speed, assuaging journalists was an inevitable feature of the launch. In May 1949, Jaguar sent a car and a party of motoring journalists to the Jabbeke highway in Belgium, a straight, well-surfaced motorway famous for speed-record attempts. Results would have supported a name-change—upward. Timed two-way tests yielded a mean maximum speed of 132.596 miles per hour, with the standing mile covered at an average of 86.434 miles per hour. To prove the XK120's tractability, the car was driven past the journalists at a modest 5 miles per hour in top gear.

The car that supposedly influenced William Lyons when designing the XK120: the prewar BMW 328 Mille Miglia.

(ABOVE) The XK120's success in rallying and races is epitomized by this period painting by Tony Smith recording the car's first major win: the British Silverstone Circuit Tourist Trophy Race in 1949.

(RIGHT) A later all-steel-bodied XK120 with the flared-in side light pods on the front wings, previously a separate chromed item. All steel-wheeled cars retained full rear-wing body spats.

1950: Metal to Metal

Jaguar could sell more cars if it could build them more quickly. Switching the XK120 to predominantly steel bodies would allow for faster assembly, and let more drivers get their hands on them. Out of a total 240 alloy-bodied cars produced, virtually all were sold abroad. The last left the production line in April 1950.

Because the entire body structure had to be reengineered for steel production, no panels were common to both cars. The entire bulkhead area was new, and although the bonnet, doors, and boot remained aluminum, they too were redesigned. The changes added around 1 hundredweight (or 112 pounds).

The first steel-bodied XK120 left the factory on April 20, 1950, and spearheaded Jaguar's attack on world

markets, not least in the United States, starting with the car's appearance at the 1950 New York Motor Show. By October, Jaguar had sold 2,000 XKs

1951–1952: Expanding the Range

A 1951 Jaguar service bulletin described enhancements that improved the XK120's performance. Higher compression ratios of 8:1 and 9:1 were possible

with better-quality fuel becoming available. High-lift camshafts and new carburetor needles were also available, along with a twin-pipe exhaust system with a straight-through silencer. These modifications could boost power from 160 brake horsepower to around 180. Other performance modifications included different axle ratios, stiffer antiroll bars, and rear springs.

There were nonmechanical enhancements too. Replacing the normal windscreen with small racing screens with cowls improved aerodynamics. Buyers could also secure lightweight racing-type bucket seats and a metal tonneau cover.

Wire wheels became a factory option in 1951. The same size and diameter as the steel wheels, they were splined to fit revised axle hubs with knock-on spinners. Where wire wheels were fitted, the rear wheel spats were completely removed.

In total, 1,114 roadsters were produced in 1951.

The Tin-Top

A fixed head coupe version of the XK120 was launched at the 1951 Geneva Motor Show, taking styling cues from the still-new Mark VII saloon. The coupe improved visibility and access with an enlarged glass area, extra headroom over the soft top, wider doors, and more space between occupants and the dashboard.

The greatly enhanced interior made the fixed head a comfortable long-legged grand tourer. Although the layout was similar to the roadster's, the instrument panel was finished in figured walnut veneer, now with an open-storage glove box area on the passenger side. Walnut veneer continued to the door cappings.

Opening quarterlights on the doors and rear side windows, and opening side vents in the front wings provided greater airflow into the cockpit. (This later feature was added to the roadster from chassis numbers 660675 (right-hand drive) and 671097 (left-hand drive).) The fixed head was also equipped with a standard heater.

Twin sun visors were now standard, as well (though still not fitted to the roadster), and there were even twin interior courtesy lights mounted in the rear quarters of the quality headlining. Other saloon-type features included glass door windows that could be wound down

The interior of the XK120 drophead and fixed head models. Extra trim, wood veneer, wind-up windows, and lockable doors—a true grand tourer.

and up, and proper internal and external door handles, making it possible to lock the car.

In 1952, there were few changes and 1,658 roadsters and 1,318 fixed heads were made.

1953: The Best of Both Worlds

Buoyant sales continued for the XK120 with a further 1,260 roadsters and 868 fixed heads sold.

Jaguar took another XK120 back to Belgium's Jabbeke highway in October 1953 for high-speed testing under controlled conditions. Although more modified than previously, the car managed to achieve an amazing 172.4 miles per hour top speed, reemphasizing the XK's dominance over other cars of the period (and many since!).

In April 1953, Jaguar announced a third variant to the range, the drophead coupe, a model just as beautiful as the other XKs but more practical. The essence of the new model was that it was

still an open two-seater but with a much more practical (and attractive) hood arrangement.

Following the fixed head the new drophead also had wider doors with glass wind-up windows with chromed frames and quarterlights. Internal trim also followed the style of the fixed head.

Externally finished in mohair, the roof also had an internal headliner covering the framework. There was even a courtesy light. The large rear screen was produced in Perspex and could be unzipped for airflow. By means of three over-center toggle catches, the roof could be unlatched, folded down to rest behind the seats, and lifted back to the closed position, all from the driver's seat. A supplied tonneau cover protected the interior when the hood was folded.

Mechanically the drophead wasn't altered, and the same year Jaguar offered a close-ratio gearbox for all models. The drophead was available as a standard model with steel wheels and as a Special Equipment model with wire

(RIGHT) The XK120 drophead provided creature comforts with open-air motoring, yet retained all the elements of the original styling. The well-designed roof could be put down and raised in seconds from the driver's seat.

(BELOW) The stylized lines of the XK120 fixed head made this the most civilized closed sports car on the market at the time. The roof was modeled after Jaguar's new Mark VII saloon launched a year earlier.

wheels and many of the other upgrades mentioned earlier.

In 1953, 1,251 dropheads were produced, along with another 515 in 1954. It was the last and least common of the XK120 variants. All XK120 models were discontinued in September 1954.

1954–1956: The XK Is Dead, Long Live the XK

Jaguar's success with the XK120 and the Mark VII saloon left the factory working at maximum capacity to meet demand. Still Lyons was looking ahead. The XK120 was five years old and not without critics, particularly in terms of interior space. In the States there had also been complaints over the vulnerability to body damage. The car was ready for a "facelift."

October 1954 saw the launch of the XK140, again at the British Motor Show. This time all three variants (roadster, fixed head, and drophead) were announced together.

The external changes were the most noticeable, and plainly brought about by demands from the North American market for a more practical design. Gone were the slimline bumper bars, replaced by a substantial ribbed type with prominent overriders (with a metal valance to hide the mountings), more suitable to the rigors of knock-for-knock parking and other minor damage. At the rear, quarter bumpers were fitted flush to the bodywork with the center area reserved for the license plate. More substantial bumpers made the XK140 3 inches longer than the XK120.

A new radiator grille made from cast metal featured fewer and thicker vertical slats, adorned with an enameled badge (versus bonnet mounting on the XK120). Chrome trim extended from the top of the radiator grille to the furthermost edge of the bonnet near the scuttle.

COMPETITION TYPE

The C-Type sports racing car, derived from the XK120 and a winner at Le Mans in the early 1950s.

At the time of the 1950 Le Mans 24-hour race, Lyons realized that most of the cars competing were neither particularly advanced nor specially designed for the race. Jaguar's own XK120 was close to those cars in performance and reliability. A modified XK120, he realized, might just win. The potential boost to Jaguar's reputation validated the effort, leading to the XK120 C (C-Type).

Engine modifications increased output to 205 brake horsepower. The brakes were fitted with automatically adjusting front shoes, the suspension was improved, and the steering box was replaced with a rack-and-pinion system.

The C-Type featured a tubular framework fitted with a more aerodynamic body crafted in aluminum. A one-piece bonnet with flared in-wings and lights was hinged from the front, allowing easy access to the whole front of the car. The rear body section was also one piece and easily unbolted for access to the rear axle and suspension.

Lyons's hunch was correct. Jaguar's C-Type topped the field at Le Mans in 1951 and again in 1953. Just 54 C-Types were produced.

Lighting was also changed. Wing-mounted sidelights were larger. Due to legislation, flashing indicator lights were mounted separately at the base of each front wing. The headlights were of a larger, more prominent style, incorporating a center metal "J" badge. Special Equipment models featured Lucas fog and spotlights mounted on the valance. At the rear, the light clusters were larger and incorporated flashing indicators, and there was another chrome strip along the center of the boot lid. An area was cast in the strip to accommodate an enameled badge boasting Jaguar's Le Mans successes.

The new face of the XK in the XK140 from 1954. Note the different side and flasher lights, radiator grille, and bumper bar treatment.

There was also a new pushbutton handle and lock for the boot lid.

The license-plate mounting now occupied the space where the spare wheel was accessed on the XK120; the XK140's spare wheel was mounted beneath a panel in the boot.

The new roadster and drophead coupe retained most of the features of the XK120, while the fixed head was a very different car in body and trim, much more practical all around. The windscreen was moved forward, providing an extra 1½ inches of headroom to the roof. The roof was extended rearward another 6¾ inches to provide space inside for "occasional" rear seating. This meant a larger window area, both in the doors and rear side windows. The doors were 5½ inches wider than those of the XK120 and were fitted with modern pushbutton door locks.

Under the skin there were significant changes from XK120 to XK140. First, the engine was now the 190-brake-horsepower Special Equipment version

as standard. Owners wanting more power could have a C-Type cylinder head fitted. A more important change was moving the engine 3 inches forward in the chassis to provide better handling balance and enable the bulkhead to be moved forward for more legroom.

A larger slanted radiator fitted with an eight-bladed fan improved engine cooling. Replacing the aged Burman recirculating-ball system with Alford & Alder rack-and-pinion steering improved handling. The change also provided a more comfortable angle for the steering column than the more vertically positioned XK120.

Although the XK140 used the same gearbox, optional Laycock de Normanville overdrive (on fourth gear) required a slight change to the chassis. A BorgWarner three-speed automatic transmission also became available.

Internally, the new models were roomier due to the 3 inches gained by moving the engine forward. In the fixed head and drophead models, the

bulkhead was swept around the engine allowing the seats to be moved forward nearly a foot. This provided space for two occasional rear seats. As the twin 6-volt batteries had previously been located in this area, they were now moved inside the front wings.

Changes to the fixed head roof created more headroom and more legroom for the driver and front-seat passenger. Wood veneer remained, and the dashboard was virtually the same except for the addition of the centrally mounted switch to control the flashing indicators; for overdrive (where fitted), the illuminated switch was mounted on the dashboard to the side of the steering column.

Although the drophead's bulkhead move was the same, this model didn't benefit from the swept area so there wasn't as much improvement in overall room. It did, however, allow for an increase in front-seat adjustment by about 3 inches. Apart from the extra rear seating, all other aspects were unchanged, although for this model and the road-

Despite the many changes to the XK140, in roadster form, it retained the styling features of the earlier car.

(**ABOVE**) "Occasional" rear seating in the XK140 drophead and fixed head allowed for the front seats to be moved farther back. (The speaker grilles below the tonneau cover were *not* a period feature!)

(**ABOVE, RIGHT**) BorgWarner supplied three-speed automatic transmissions for the XK140, operated from this Bakelite lever set into the dashboard.

(**MIDDLE**) The XK140 fixed head with its longer roof, larger side glass area, and wider doors incorporating pushbutton door handles. As with all steel-wheeled XKs, full spats were fitted as standard.

(**BOTTOM**) Comparison of styling features of the XK120 fixed head (left) and XK140 fixed head (right). Despite the slightly lower roofline over the rear screen, the larger side window area is clear. The bumper bars provided better protection against incidental damage.

(**ABOVE**) Viewed from the side, the XK140 drophead was remarkably similar to the previous model.

(**RIGHT**) Launch picture of the XK150 fixed head, seen in US specification with the white indicator lenses and revised headlights. The auxiliary Lucas lights were an option. Note the wider grille to aid cooling of the XK engine bay.

ster the battery was changed to a single 12-volt unit (sited in one front wing).

The roadster benefited from the same engine move and the 3 inches extra adjustment in the seating. By lifting the car's scuttle line 1 inch, the steering column was raised sufficiently to provide more space between the base of the steering wheel and the driver's legs. As the roadster did not have the extra rear seating, this left slightly more space for luggage. With the bulkhead changes, the dashboard was more upright but retained the leather covering of the earlier model.

The XK140 range continued in production until the end of 1956, with the very last cars coming off the line as late as February 1957, by which time 8,676 cars had been produced.

1957: 140 Out, 150 In

Announced in May 1957, the XK150 was not so much a new design as further development of the XK140. Initially, it was available only in fixed head and drophead form.

The chassis was basically the same; even the XK140 suspension was carried over with minor changes. The rack and pinion also remained, but with rubber damping to eliminate kickback.

Even the engine was the same 3.4-liter XK unit available initially in two options: the standard, devel-

oping 190 brake horsepower, or the 210-brake-horsepower Special Equipment version. The latter had a new cylinder head (the B-Type) providing improved performance down the rev range, achieved with larger exhaust valves retaining the smaller-diameter inlet port throat of the standard head. The result was the same power as the old C-Type head but at lower revs. The engine was fitted with a separate water gallery instead of an integral type.

AN ALPHABETICAL MOVE: C TO D

Jaguar's C-Type demanded an encore.

In design and style, the D-Type was a much different car. Designed by aerodynamicist Malcolm Sayer, it was very aerodynamic and lightweight. It featured a center section of pure monocoque (chassis-less) construction and a front bulkhead strengthened by a transverse box section. Frames from the front tubular part of the car carried the engine and suspension. The center section was riveted aluminum paneling strengthened with internal sills front to back. At the rear, an unstressed tail section was bolted to a double-skinned bulkhead.

The front suspension was similar to the road-going XKs with ride-height adjustment. The rear suspension was via a bottom torsion bar anchored at its center onto the rear bulkhead and connected to the rear axle by a steel-plate arm.

Rack-and-pinion steering followed the XK140 design, and braking was an improved version of the Dunlop C-Type system, with peg drive–mounted light alloy Dunlop wheels with a steel center section.

The XK engine was modified with dry-sump lubrication, lowering the engine and bonnet line. Weber carburetion and modifications to other ancillaries much improved performance and handling.

Although it didn't win its first Le Mans outing in 1954, under continuous modification, the D-Type achieved incredible success in subsequent years, winning in 1955 and 1956, and dominating in 1957, where it took first, second, third, fourth, and sixth positions. Just 62 D-Types were produced.

When it retired the D-Type, Jaguar decided to turn remaining monocoques at the factory into road-going sports cars to be called the XKSS. Essentially a pure D-Type that featured a modicum of practicality with a single-piece curved windscreen, opening doors, a full exhaust system emitting from the side of the car, a luggage rack at the back, and a simple hood arrangement to keep out bad weather. Just 16 were produced before a major fire at the Jaguar factory in 1957 destroyed the remaining bodies. Sold to private individuals, the rare cars are now highly valued.

the XK-SS two-seater super-sport
Available in limited quantities

Here, for the experienced sports car driver, is the translation of Jaguar experience in recent competition models into the broader category of the dual-purpose sports-racing machine. A high performance sports-touring car in the tradition of the great continental marques, the XK-SS embodies the race-proved qualities of performance, roadability, reliability and safety that Jaguar believes are as desirable on the highway as they are necessary on the track.

262 H.P. "XK" engine, 3 Weber dual carburetors, four-wheel Dunlop disc brakes, all-weather top, chrome luggage rack, full touring equipment.

Following the D-Type's splendid racing career, Jaguar converted remaining bodies into the road-version D-Type, called the XKSS.

The even more successful racing D-Type, produced from 1954 until 1957, won Le Mans on three separate occasions.

This enameled badge appeared on XK models, updated each year to indicate Jaguar's Le Mans successes.

Automatic transmission XK150s were fitted with this slide quadrant lever, suitable to left- and right-hand-drive cars.

Comparison of frontal aspects of the XK140 drophead (right) and the facelifted XK150 drophead (left) clearly shows the many changes, not least the curved one-piece windscreen.

The transmission was also the same Moss type, also available with overdrive, or the BorgWarner automatic. The quadrant lever to control the automatic gear selection took the form of a Bakelite panel below the center dashboard area, eliminating the need to change position for left- or right-hand drive models.

The single biggest mechanical change to the XK150 came with the braking system. Although catalogued as available with the old Lockheed drum system, all XK150s were fitted with Dunlop 12-inch disc brakes at each wheel with Lockheed servo assistance. Developed from Jaguar's racing exploits, the system was far more efficient than the old drums, although Jaguar was later than other manufacturers in fitting them.

In styling terms the XK150 retained a lot of the original Lyons design but with smoothed sides and many detail changes. Conventionally produced in steel, now

The XK150's interior was more modern, with no woodwork to be found. Note the slight hint of crash safety with the padded roll on the dashboard and the semiflush slide door handles.

The XK150 fixed head with even more glass area and a signature feature of all XK150s: the wraparound single-piece windscreen.

Even up to the end of XK production, steel-wheeled cars were still available and fitted with rear wing spats. This is the pretty XK150 roadster with the top erect—a much more practical affair than on previous XKs.

with only the bonnet and boot lid in aluminum, the car looked more slab-sided with a higher waistline.

The front bore a "corporate" design following the style of Jaguar's then new 2.4-/3.4-liter (Mark 1) saloons with a wider, thin-slated radiator grille to improve cooling (still fitted with an enameled badge). A chromed leaping Jaguar could be specified, fitted on the bonnet behind the grille.

The substantial bumper bar first seen on the XK140 was restyled with a curved indentation in the center to accept the larger radiator grille. Lighting followed XK140 practice. The rear now featured a wraparound one-piece bumper bar (like the saloons) and due to the reshaping of the rear wings, larger plinths were required for the light units. More chrome was added to the boot lid and, of course,

the enameled badging was amended to include the latest Le Mans wins.

Apart from the obvious visual changes to the body style, a major change came with the windscreen: For the first time, a single-piece curved screen was fitted on a higher scuttle.

As before, XK150s were available with a choice of steel or wire wheels; the former retained full spats over the rear wheels.

The interior came in for a significant makeover. Out went the wood veneer in favor of padded leather, both for the dashboard (the main instrument cluster area in a contrasting color to the other trim) and the door cappings. Instrument layout was more akin to the small Mark 1 saloons and the rearview mirror was relocated to the roof area, rather than on the dashboard top rail. Although

much of the switchgear was the same, the flashing indicator switch was now a "modern" stalk on the steering column. Heating and ventilation followed XK140 practice.

The front seats were of a similar design but slightly wider, folding forward for ease of access to the still occasional rear seating. A hinged lid in the rear compartment provided access to the boot area for carrying lengthy items like golf clubs.

The drophead coupe had a hood and mechanism very similar to but longer than the XK140s. Although working in the same speedy manner, when down it stood higher and with the fitted tonneau cover, looked a little unsightly and restricted rearward visibility.

The XK150 drophead with its more slab-sided styling and prominent hood. For the first time, the XK150 roadster offered wind-up windows and lockable doors. The higher waistline suited this model.

1958: Increasing the Performance and the Range

In March 1958, Jaguar introduced a new but not unexpected model to the XK150 range, the roadster. As with the previous roadsters, this was strictly a two-seater. A new rear panel extended into the cockpit to finish just behind the front seats, creating a long and wide deck ahead of the boot lid. There was just enough space behind the seats to accommodate a new (and more attractive) folding roof. When down, the roof was secured with straps and a neat cover snapped in place to hide it.

Glass wind-down windows in the doors with chromed surrounds provided a more civilized approach than floppy side screens. There was even a chrome-plated finisher along the top edge of each door, the door tops themselves sculptured with a gentle curvature to the horizontal line. Overall there is no doubt that the XK150 roadster was the most attractive of the XK two-seaters even if the bodywork looked a little top-heavy compared to the earlier cars.

At the time, the roadster was introduced, a more powerful option known as the S model became avail-

able. Improved performance came from a new straight-port cylinder head; straightening the ports produced better flow of the air-fuel mixture at higher revs. The S engine also included a triple 2-inch SU carburetor setup with revised manifold. Stronger clutch assembly and lightened flywheel were also incorporated. This new engine was rated at 250 brake horsepower. An improved braking system incorporated quick-change square disc pads.

When the S engine was fitted, a chromed stylized "S" also appeared on the leading edge of each door.

The XK150 3.8-liter S engine with triple 2-inch SU carburetors.

Comparisons showing the styling changes over the years in the roadster models, from the XK120 through the XK140 and onto the XK150.

The S model was available only with the manual gearbox and if overdrive was fitted, the operating switch was repositioned to the transmission tunnel area.

Initially the XK150S roadster was available only for overseas markets and remained so until October 1958.

1959–1961: More Performance and a Wider Range

Early in 1959, the S model became available for the fixed head and drop-head models. As many competitors' cars were now being equipped with disc brakes by now, these were standardized on all models.

For 1960, the XK150 range was expanded again by the fitting of another even more powerful variant of the XK engine, this with a capacity of 3,781cc (3.8 liters). This engine was not only available in all three body styles, but also with the conventional B-Type cylinder head and twin carburetors, or with the straight-port head and a triple-carburetor S spec setup. The new engine was rated at 265 brake horsepower in S form with triple SU 2-inch carburetors. The bores were increased from 83 to 87 millimeters and now fitted with liners and the water passages modified to aid cooling.

XK150 production finished in January 1961 with a total of 9,385 cars. Combined with XK120 and XK140 totals, Jaguar produced over 30,000 XKs in just over twelve years. An even more remarkable car would follow the XK line.

SPECIFICATIONS

MODEL	XK120 Standard	XK120 SE	XK140 Standard	XK140 SE	XK150 3.4
ENGINE SIZE	3,442cc	3,442cc	3.442cc	3,442cc	3,442cc
CARBURETION	2 x 1.75 SU	2 x 1.75 SU	2 x 1.75 SU	2 x 1.75 SU	2 x 1.75 SE
MAXIMUM BHP	160@5,000	180@5,300	190@5,500	210@5,750	190@5,500
MAXIMUM TORQUE	195@2,500	205@4,000	210@2,500	213@4,000	210@2,500
GEARBOX	4-speed	4-speed	4-speed	4-speed	4-speed
AUTOMATIC	n/a	n/a	BW 3-speed	BW 3-speed	BW 3-speed
0 TO 60 MPH	11 sec.	10 sec.	10 sec.	8.4 sec.	10.5 sec.
STANDING ¼ MILE	18 sec.	17 sec.	17 sec.	16.6 sec.	17.5 sec.
TOP SPEED	120 mph	120 mph	120 mph	120 mph	120 mph
AVERAGE FUEL CONSUMPTION	20 mpg	20 mpg	18 mpg	18 mpg	20 mpg

MODEL	XK150 3.4 SE	XK150 3.4 S	XK150 3.8	XK150 3.8 S
ENGINE SIZE	3,442cc	3,442cc	3,781cc	3,781cc
CARBURETION	2 x 1.75 SU	3 x 2 SU	2 x 1.75 SU	3 x 2 SU
MAXIMUM BHP	210@5,750	250@5,500	220@5,500	265@5,500
MAXIMUM TORQUE	216@3,000	240@4,500	240@3,000	260@4,000
GEARBOX	4-speed	4-speed	4-speed	4-speed
AUTOMATIC	BW 3-speed	BW 3-speed	BW 3-speed	BW 3-speed
0 TO 60 MPH	8.5 sec.	7.3 sec.	7.6 sec.	7 sec.
STANDING ¼ MILE	16.9 sec.	15.1 sec.	16 sec.	15 sec.
TOP SPEED	120 mph	136 mph	136 mph	136 mph
AVERAGE FUEL CONSUMPTION	20 mpg	16 mpg	15 mpg	13 mpg

CHAPTER FOUR

THE FLAGSHIP SALOONS

Mark VII to 420G

From the time of the SS Jaguars in the 1930s, William Lyons's ultimate ambition was to build a luxury 100-mile-per-hour saloon that would match anything else on offer.

The process was well underway by 1948, when a new chassis with independent front suspension appeared in the Mark V saloons. Another huge, contemporaneous, milestone was the XK twin-camshaft six-cylinder engine. Launched in the low-production XK120 sports car, this power plant was conceived to further Lyons's envisioned 100-miles-per-hour saloon. Both engine and chassis proved their worth and Lyons's desired saloon followed in 1950.

Development of the high-performance saloon led Jaguar to significant upgrades over the years and new technologies in the 1960s, making Jaguar's top-of-the-range models the benchmark for other manufacturers.

1950: The Prima Ballerina Arrives

The Mark VII was critical to Jaguar's success as a prestige saloon manufacturer—as important as the XK120 had been to its sports car business.

To complete the package for this saloon, Jaguar needed an all-new body design to accompany the chassis and engine. The design would need to suit the car's advanced performance and satisfy buyers loyal to Jaguar. Equally important, it would need to entice a wider, overseas, audience in markets like the United States.

Lyons's unleashed his creative flare, working in his usual way with full-sized wood mock-ups. Headlights were fully set within the bodywork with matching flush-mounted auxiliary lighting

beneath. The radiator grille was more restrained than previous offerings, without a pronounced surround or badging. The double-bumper bar arrangement from the Mark V was simplified as a single pressing for the Mark VII with new overriders.

The XK engine sat beneath a single-piece alligator-style bonnet hinged at the rear and incorporating side chrome trim. A center chrome strip at the front led to an integrated jaguar-head emblem and winged badge with the Jaguar name. Overlooking the bonnet was a surprise two-piece windscreen, an approach more common on prewar cars.

The sideview featured a flowing line running from the top of the front wings down through the doors and lifting over the rear wheels (a styling touch borrowed from the XK120). Rear wheels were fully enclosed with fitted spats contoured to the bodywork. With no running boards, the central body section was widened with flush-fitting sills integrated into the body. All four doors were now forward-hinged in the modern style.

At the back, the boot had a shallower slope and incorporated an attractive rear license plate nacelle with twin operating handles. Bumper bar treatment followed the style at the front, and rear lighting was minimalist.

Less brightwork adorned the Mark VII than its predecessor. The chromed brass window frames were retained, as were the waistline chrome strips running from the leading edge of the bonnet through the doors to neat "spears" on the rear wings. Jaguar's new-style pushbutton door handles and locks (first seen on the Mark V) were also retained, as was a steel sliding sunroof.

SIXES AND SEVENS

Jaguar's design vision was clearer than its naming conventions. Their first postwar model was the Mark V, yet there had been no official Mark IV—or III or II or I! Mark V was chosen because it was the fifth prototype of the model, information unknown to the buying public. Release of the Mark V didn't clarify things for those expecting a Mark VI to follow. The new car went by Mark VII.

But that was Bentley's fault, as the competitor had already introduced a Mark VI saloon in 1946. The companies agreed that competing models with the same name served neither maker, so Lyons agreed to skip the number!

(ABOVE) The Mark VII introduced a new era for Jaguar in the 1950s. Here a very early example is shown in the foreground, behind which is the final interpretation of the styling, the Mark IX, the last of which were produced in 1961.

(BELOW) The beautifully curvaceous styling of the Mark VII is emphasized in this artist's impression prepared for the original press images and brochure.

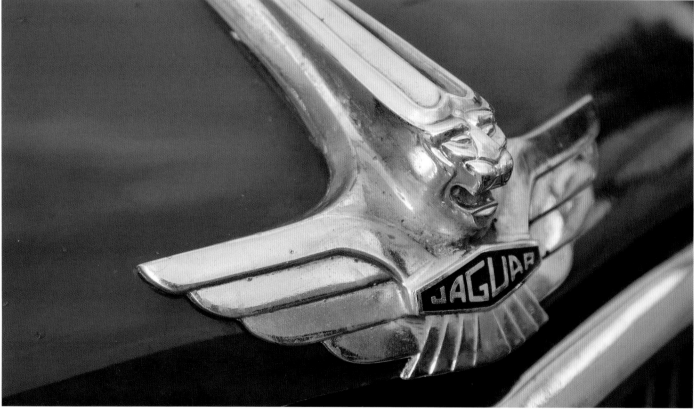

A Mark VII frontal view shows the less formal grille treatment than the previous model and the neat flush-mounted auxiliary lighting. The split windscreen was a throwback to prewar car design. Inset: The "stub" mascot head on the leading edge of the Mark VII bonnet.

The Mark VII was a large car, even by US standards, and lacked added brightwork. The somewhat slab-sided styling and large paneling suited darker colors, which were far more popular with customers than the likes of cream.

The car sat on the same sized 16-inch rims, painted to body color with chromed hubcaps also finished to body color.

The Mark VII's styling withstood the test of time, remaining in production in its original body form for eleven years.

The all-enveloping body style meant fewer, larger panels, so production was outsourced to the British car industry's major body supplier: Pressed Steel Company. Tooling up for such a large, complex body took time, which was the primary reason the Mark VII was not launched until 1950.

To accept the XK engine, the chassis was 3 inches longer than the Mark V chassis built for the old pushrod unit. Yet the XK engine sat 5 inches farther forward in the chassis, providing extra accommodation inside the car.

There were few other mechanical changes except for the braking system. An entirely new Girling design used two self-adjusting trailing shoes in the front drums and a Clayton Dewandre hydraulic servo activated from vacuum taken from the inlet manifold. Rear brake drums and design remained the same as the previous model, but the handbrake lever was relocated from under the dashboard to a conventional pull handle between the front seats.

The engine's repositioning required adjustments to the steering. Its additional weight called for strengthened suspension mountings.

Internal space in the engine bay was marginally better than the Mark V and the battery and heater boxes were mounted on the bulkhead. Manifolding similar to the XK120 was adopted.

Taking advantage of the increased width and extra space gained by repositioning the engine, the Mark VII's interior was exceptionally well appointed and roomy. Leather seating reverted to a pleated style with a large rear bench mounted slightly forward of the wheel arches, with a center folding armrest. The front individual seats were wider than the Mark V's and mounted slightly forward, providing an extra 3 inches of legroom for rear passengers. All occupants gained an extra 2 inches

(**ABOVE**) One of the two compre-hensive toolkits fitted in the front door cavities on all Mark VII to IX models.

(**ABOVE, RIGHT**) The Mark VII's impressive interior with extensive veneer, a full complement of instruments, and the original trademark four-spoke steering wheel. Note the standard fitment sunroof, the inset sun visors, and trigger-style door lock/handles.

(**BELOW & RIGHT**) Period *Autocar* magazine front cover after the launch of the Mark VII at the Waldorf Astoria Hotel in New York, with Jaguar's press photograph at the same venue. (Note the magazine had inverted the image to suit their front cover. As a result, the front of the car appeared to the right edge of the page instead of against the spine!)

of headroom over the previous model. The door panels were well appointed, retaining the trigger-action door handles and pulls from the Mark V.

Walnut veneer was retained extensively with a new dashboard. The instruments and switchgear were mounted in a similar manner and style to the outgoing model, although the instrument design was amended with more traditional needle pointers and clearer legends. An 18-inch Bluemels four-spoke steering wheel was also retained as a trademark feature. The standard-equipment heater was supplemented by opening ventilators set into each front wing to aid flow of cool air in hot climates.

An extensive range of tools was provided, fitted in fold-down boxes in each front-door cavity.

The boot area was the largest ever seen on an SS or Jaguar model. Internally, over 4 feet in length and 2½ feet in depth, and nearly 4 feet at its widest point, it was hindered only by the upright-mounted spare wheel, designed to be extracted without affecting luggage. The fuel tanks were fitted inside the rear wings out of the way. The heavy rear-hinged boot lid was supported by a telescopic stay.

The Mark VII launched at the 1950 British Motor Show at Earls Court in London. A single example in Light Metallic Blue featured on a revolving plinth stole the show. One journalist from Britain's *Autocar* magazine christened the car "The Prima Ballerina."

With promises of 100-mile-per-hour performance from its XK engine, superb handling from its independent-front-suspension chassis, excellent safety from front self-adjusting brakes, and the opulent interior of walnut and leather, it was to be a sure-fired success. The greatest surprise for visitors, however, was the price. At just £988 before tax (same as

the outgoing Mark V saloon and the XK120), it fell below the UK's top tax bracket and assured its competitiveness.

With all expectations achieved, the next stop for the new model was the US launch at the Waldorf Astoria Hotel in New York. The car won maximum acclaim and sent Jaguar back across the Atlantic with over $30 million in orders.

1951–1952: Moving Forward

Having outgrown its Foleshill factory, Jaguar in 1951 and 1952 moved to the ex–World War II Shadow factory run by Daimler in Browns Lane near Allesley, just a few miles away.

The Mark VII was selling well. It came in for various revisions during this time, many to satisfy the needs of overseas markets. For colder climates, the cylinder block was modified to take a standard-sized water heater element providing quicker warmups. For hotter climates the original aluminum cooling fan was replaced with a steel unit running at higher speeds. Bodily, the Mark VII gained improved sealing for its doors and windows.

A two-speed wiper motor was fitted in 1952 along with the standard windscreen washer system. Uprated Girling dampers also were fitted that year with the option for owners to upgrade their earlier cars. Wheel rims increased in width from 5 to 5½ inches, and gearboxes were shortened, requiring a different propshaft.

1953: Automation

The US market was vital to Jaguar. It was also accustomed to automatic transmissions. Long suffering the slow and tiresome gear changes inherent with the old Moss four-speed manual gearbox used in theirs and other British cars, Jaguar considered it important to offer an automatic transmission version

EXTRA PERFORMANCE

The standard engine was rated at 160 brake horsepower. Fitting some of the modifications identified for the XK120 bumper power to 180 brake horsepower or even higher with the C-Type cylinder head. A close-ratio gearbox, uprated dampers, and stiffer springs would improve handling to match the increased performance. Jaguar even produced a heavyweight sump guard for those who went rallying!

of the Mark VII that they felt could double US sales.

Jaguar chose the US-built BorgWarner as the most suitable automatic gearbox for the Jaguar engine and, due to increased popularity, BW would soon set up a manufacturing facility in the UK. Jaguar opted for the BorgWarner DG three-speed with a "kickdown" facility enabling the driver to select a lower gear at full throttle. Gear selection was via a quadrant in Bakelite mounted above the steering wheel and set into the top of the dashboard. The Mark VII retained the existing 4.27:1 rear-axle ratio.

The automatic transmission changed front seating from the individual type to a single full-width bench with some walnut veneer capping set into the upholstery at the rear. The handbrake reverted to an umbrella type under the dashboard near the driver's door.

Minor mechanical changes in 1953 included a revised cooling system with yet another new fan, a new water pump, an enlarged bypass hose, modified induction manifolding, and a new radiator

The Mark VII became available with automatic transmission in 1953. With this came an optional full-width bench front seat with simple backing incorporating a veneered panel and ashtray.

The BorgWarner automatic transmission selector was this Bakelite lever and surround set into the dashboard area of the Mark VII to IX models.

with an expansion chamber on top of the header tank.

The antiroll bar was altered to accommodate a pressed-steel engine oil sump (replacing the cast-iron type). Engine valve springs were altered, and a second throttle-return spring was fitted. Finally, telescopic shock absorbers replaced the aged lever-arm type on the rear suspension.

By the end of 1953, nearly 17,000 Mark VIIs had been produced and sold.

1954–1955: Relaxed Driving and a Facelift

The more refined BorgWarner automatic transmission motivated Jaguar to improve the manual gearbox model. Although the conventional four-speed provided excellent acceleration and good driver control, on long-legged highway journeys it revved the engine pretty hard, which didn't help fuel consumption, engine wear, or driver comfort.

In January 1954, Jaguar announced the availability of a Laycock de Normanville overdrive, operational on fourth gear only. It used an epicycle gear attached to the output shaft of the Moss gearbox, controlled via a cone clutch from a switch on the dashboard. Overdrive of course allowed the engine to rev less frantically at high speeds. A lower rear-axle ratio of 4.55:1 provided better top-gear acceleration.

By 1954, Jaguar had produced 20,937 Mark VIIs, and after four years it was due for a facelift. Although not rebadged, in September Jaguar introduced the Mark VIIM. Bodily the same, the trim changed, much in line with the XK140 sports. The front and rear bumper bars were of a simpler design with taller overriders. The rear bumper received a pronounced wraparound the rear wings for better protection.

The tripod headlights were replaced with the Lucas J700s. Out, too, went

The Mark VIIM introduced in 1954 was fitted with simpler bumper bars (following Jaguar's other models from the XK140 on), separate auxiliary lighting, and flashing indicator lenses. By this time, Jaguar had abandoned painting the hubcap centers to match body colors.

A rear view of the Mark VIIM shows the different lighting and lack of badging. Also note the simpler styling of the bumper bar, which now wraps around to protect the rear wings from damage.

the old trafficator indicators, replaced with front wing–mounted flashing lights, while standard Lucas types replaced the flush-mounted auxiliary lights, mounted externally on the front valance. The space in the front wings for the older lights was now taken by chromed horn grilles. At the rear, larger light units and plinths incorporated twin-filament bulbs (for the flashing indicators) and built-in reflectors. The wings were altered to accommodate the lighting changes front and rear.

The engine was fitted with ⅜-inch camshafts, which boosted output to 190 brake horsepower. Gearbox ratios were also altered, and suspension was uprated with larger-diameter torsion bars reducing roll and improved ride height.

Internally, the Mark VIIM was similar to the Mark VII. Dunlopillo now supplied the foam fillings to the seats and the center steering boss was flat rather than domed (with a revised jaguar head).

1956: Seven into Eight Will Go

Sales of Mark VIIMs continued through 1956 without change, but in October Jaguar announced a new additional model: the Mark VIII. Not an entirely new car, it was another facelift, though one destined to develop further.

Although it had the same body as the Mark VII, the Mark VIII gained significant enhanced trim changes. At the front, Jaguar returned to a more prominent radiator grille with a pro-

nounced chromed surround at the top incorporating a winged badge with the Jaguar name. The chromed center bonnet strip was shorter to accommodate a large chromed leaping Jaguar mascot. A most significant change came with the final replacement of the split windscreen with a single-piece screen with matching chromed-trim surround. This necessitated a change to the wiper equipment.

The somewhat slab-sided styling of the earlier car was disguised by a chromed strip that followed the con-

tours of the body from the leading edge of the front wheel arch down through the doors to the rear-wheel spats, continuing horizontally to finish at the rear valance. The spats incorporated a cutaway area at the bottom to reveal part of the wheel.

With the new side chrome trims, the Mark VIII could be ordered in a wider choice of color schemes, many of which were duotone finishes with the lighter color below the trim, the darker color above, including the bonnet, boot, and

(ABOVE) Mark VIII and IX models featured the revised radiator grille with large leaping Jaguar mascots.

(ABOVE, RIGHT) The luxuriously fitted rear compartment of the Mark VIII and IX with veneered occasional tables, multiple cigarette lighters and ashtrays, and a split rear seatback arrangement. These models were also fitted with a clip-on, long-pile nylon overrug for extra rear passenger comfort.

(RIGHT) The Mark VIII introduced a new more upright dashboard arrangement, minor revisions to instruments and other surrounding trim, and relocation of the ratchet handbrake from under the dashboard to the floor between the seats, unless a bench seat was fitted.

(OPPOSITE) Jaguar's Mark VIII heralded more brightwork, enabling duotone exterior paint finishes. Out went the split windscreen and in came a more prominent radiator grille.

roof areas. This modernized the look of the car, making it appear longer and lower.

The interior also received lots of minor improvements. Dunlopillo seat fillings were increased and the front-seat height altered slightly. Both bucket and bench front-seat backs incorporated walnut veneered fold-down occasional tables set into veneered surrounds. For bench-seat-equipped cars the rear also featured a veneered magazine rack and clock. The rear seat was also much more luxurious, styled to resemble individual seating.

Front door trims now incorporated map pockets and all four doors had ashtrays. To match, there were cigarette lighters in each B/C post, set into the wood veneer, and another in the dashboard for front-seat occupants. The wood veneer paneling was amended, the door cappings no longer incorporated ashtrays, and the dashboard top rail was of a smoother design. The dashboard

With the Mark VIII came more power from the 3.4-liter XK engine along with other more minor mechanical revisions. More power also meant a dual exhaust system.

was very similar but more vertical with a revised instrument layout.

The boot interior was fully trimmed with Hardura, including the spare wheel and the underside of the boot lid.

The Mark VIII's engine was fitted with Jaguar's B-Type cylinder head as standard. With a revised inlet manifold with separate bolt-on water rail, the latest SU carburetors, and a twin exhaust system, the Mark VIII had a useful extra turn of speed and improved top-gear performance.

Both manual and automatic transmission models were available, and the latter now included a unique Jaguar feature: Intermediate Speed Hold. By means of a dashboard-mounted switch, the driver could select or hold an intermediate gear for hill-climbing and more spirited driving style. This became a standard feature on all subsequent automatic transmission Jaguars for many years.

In 1956, both the Mark VIIM and Mark VIII were listed as current models to help move unsold examples of the earlier car.

1957–1958: Eight Becomes Nine

Mark VII production finally came to an end early in 1957 and the Mark VIII was destined to come to the end of its short production life in 1959. Along the way there were still updates and modifications.

In 1957, the engine received drilled camshafts for quieter cold starting. Nylon interleaving to the rear springs was fitted from 1958 and the exhaust pipe diameters were reduced.

In 1958, under pressure from the US market, Mark VIIIs to order were supplied with power-assisted steering by Burman.

Just 6,185 Mark VIIIs were produced during 1957 and 1958. Although listed alongside another new model, the Mark VIII was available only until 1959 to help clear stock and satisfy minimal demand. That new model, the Mark IX, looked exactly like its predecessor in nearly every way—the only way to discern a Mark IX from a Mark VIII was by the boot-mounted badge!

Most of the changes were under the skin. The Mark IX was the first Jaguar saloon fitted with the 3.8-liter version of the XK engine (first seen in the XK150 sports car). Additionally, it was the first Jaguar saloon fitted with power-assisted steering. All-round disc brakes were standard.

Details and benefits of the 3.8-liter engine are covered in an earlier chapter. Suffice to say the increased performance to 220 brake horsepower helped keep the Jaguar saloon competitive in world markets. The same four-speed gearbox (with or without overdrive) and BorgWarner automatic transmission options were available.

The Burman power-assisted steering system employed a Hobourn-Eaton eccentric-rotor pump driven from the rear of the dynamo shaft, providing a continuous flow of oil through the hydraulically assisted worm and recirculating ball (larger) steering box. A large circular oil reservoir was sited in the engine bay.

The Dunlop brakes used 12-inch discs fitted with square "quick-change" pads and twin cylinders. Lockheed provided a large suspended vacuum servo for power assistance, incorporating a vacuum reservoir to hold pressure should the engine fail. Jaguar reverted to an umbrella handbrake for all cars, even those with individual front seats.

With all these changes, the Mark IX was a more agile car that kept pace with what little competition remained in this market.

Internally, it was much the same as the Mark VIII with minor tweaks for modernity. The indicator switch was moved from the steering wheel boss to a stalk off the column. The tachometer and speedometer swapped positions, with the needles now traveling in the same direction. The heater unit was now centrally mounted on the engine bay

(ABOVE) The launch of the Mark IX, the final derivative of the Mark VII design, featuring all the changes identified for the Mark VIII, but with substantial performance upgrades.

(BOTTOM) Even though duotone paint finishes were available for the Mark VIII/IX models at no extra cost, substantial numbers were also produced in a range of single-tone schemes. The single noticeable external difference between a Mark VIII and a Mark IX is the "Mk IX" badging on the boot lid!

A Mark IX engine bay showing the 3.8-liter engine, power steering reservoir (mounted on the right), revised sighting of the heater box on the bulkhead, and the necessary twin batteries sitting astride the steering column on this left-hand-drive model.

The revised interior for the Mark IX. The speedometer/rev counter needles have changed direction, a handbrake warning light has been added to the top dash rail, the center-boss-mounted indicator switch has moved to a stalk on the column, and the handbrake has returned to the ratchet type under the dashboard, regardless of seating style.

MARK VIIIB

During the Mark IX's production, Jaguar produced a small quantity of limousine versions for a specialist market. Externally, the Mark VIIIB looked exactly like the Mark VIII/IX, and all were produced in single-tone colors (normally black). They were fitted with a revised chrome bonnet center trim to incorporate a mast for mounting a flag.

Mechanically the cars were based on the Mark VIII with a 3.4-liter engine, usually of a lower (7:1 or 8:1) compression ratio. Most were fitted with a revised pancake air cleaner, which would also be used on the Jaguar Mark 2 and Daimler Majestic Major, to be discussed in later chapters. Most had a manual transmission, some without overdrive.

Internally, Mark VIIIBs were fitted with a center glass division with suitable amendments to the styling of the rear occasional tables. Instead of the magazine rack, a cabinet with opening doors was fitted (with the clock still mounted above). Because of the center division, all these cars were fitted with a fixed bench front seat with no height or fore/after adjustment. Where manual transmission was fitted, a cutout area in the bench seat accommodated the gear lever. The rest of the interior trim was straight out of a Mark VIII.

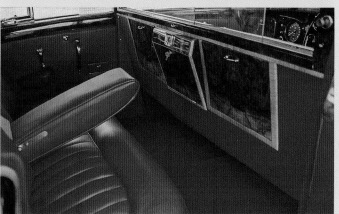

Interior layout of the rare Mark VIIIB limousines showing the cut-away front seating to accommodate the gear lever, the lack of map pockets in the front door trims, and the glass division and revised wood areas at the rear.

bulkhead and its output was increased by about 1½ kilowatts.

In the rear compartment, where a front bench seat was fitted, the centrally mounted magazine rack was changed to a shallower design with a veneered lockable lid.

1959–1961: The End of the Chassis Era

The last 62 examples of the Mark VIII left the factory in 1959 when the Mark IX took over completely. The successor bore few changes except for an improved handbrake, larger-diameter ball joints, and a tach change from cable drive to electric.

In 1960, the dashboard was fitted with a brake warning light that illuminated when the handbrake was applied or if the brake reservoir lacked fluid. Due to changes in legislation, larger rear-light units and chromed plinths were fitted, incorporating separate flashing indicator bulbs and lenses.

The Mark IX continued in production without further changes, gradually decreasing from a high of 4,915 units in 1959 to 3,555 in 1960 and 603 in 1961 (including Mark VIIIBs).

Total Mark VII to IX production amounted to just over 46,240 cars from 1950 through 1961. It was the only saloon produced by Jaguar from 1950 through 1956 and it was successful in seeing off many of its respected competitors, like Armstrong Siddeley and even Daimler. Despite its huge size and weight, it performed admirably and even in its later years could still outperform many more modern cars. But it had to make way for the swinging sixties and a new era.

1961: A New Era

By 1961, it was time for an entirely new flagship model. The Mark VII to IX

MARK VII IN COMPETITION

XXIIme RALLYE AUTOMOBILE MONTE-CARLO
JANVIER 1952

Jaguar's success in racing isn't all down to the sports cars—the saloons also played their part, and none less than the Mark VII. In rallying, examples finished fourth and sixth in the 1952 Monte Carlo Rally, and a very private entry won its class in the 1953 RAC Rally. The following years were also good, culminating in 1956 with Ronnie Adams winning the Monte Carlo Rally outright.

In 1952, Stirling Moss won the Silverstone Production Car Race, repeating in 1953 and finishing third in 1954 behind the Mark VIIs of Ian Appleyard and Tony Rolt. The Silverstone Mark VII success continued into 1955 with first through third finishes yet again.

range had lasted well, but with competitors introducing exciting new models and technologies, Jaguar had to respond.

The Jaguar Mark X (code named Zenith) was a very different car than its predecessor. Jaguar had learned a lot since commencing monocoque (chassis-less) construction with the compact Mark 1. Zenith was to be the largest, most complex and rigid construction yet attempted by Jaguar, with the help of the Pressed Steel Company,

the provider of all Jaguar bodyshells. The Mark X shell was so big it had to be built at Pressed Steel's Wiltshire factory, not their usual Midlands location.

The styling was pure Lyons, a thoroughly modern low-slung design that handled its enormous size well. There were no undulating curves, yet it wasn't slab-sided either. The bulbous body sides contributed to the overall width of the car and increased the interior accommodation considerably.

The old-school 1950s flagship saloon on the left and the new kid on the block: a 1960s Mark X. Despite the low line of the Mark X's bodywork, ground clearance is virtually the same, visibility is better and, perhaps most surprisingly, the same straight-six XK engine fits under the new bonnet!

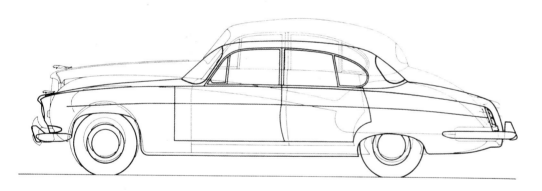

This drawing emphasizes the extra length and the low-slung styling of the Mark X in comparison to the Mark IX.

From the side, the car's length was emphasized by the low roof, and the doors still incorporated Jaguar's trademark chromed window surrounds with unique rear quarterlights. There was a distinct lack of other brightwork except for new-style door handles.

The large rear wings incorporated twin fuel tanks and the enormous boot lid opened to a cavernous, fully lined interior with even more space than that of the previous model. The spare wheel was still mounted vertically and covered; now space was conveniently situated at the side of the spare wheel for the jack and the toolkit.

Externally, new lighting units included a boot-lid-mounted nacelle. The rear body valance was more visible with the slimline bumper bar, under which the twin exhaust pipes with chrome finishers exited centrally but separately.

Overall it was a very clean design. The car sat on 14-inch rims. Although the shorter rim was unique to any Jaguar

A mid-height swage line broke up the large paneling, which helped take away from the car's general bulk.

At the front, a forward leaning, squat radiator grille, and four-headlight treatment gave a new look. With no auxiliary lighting, the same horn grilles from the Mark IX were fitted, along with new sidelights/indicators that wrapped around the sides of the wings. Slim-line bumper bars with forward-sloping overriders were fitted front and rear, the rear with a considerable wraparound to protect the wings from damage. The enormous front-hinged bonnet incorporated the grille and inner headlights, retaining a center chromed trim but with a smaller leaping-Jaguar mascot. A one-piece curved windscreen provided good visibility.

The modernity of the Mark X body style allowed a lower roofline and greater visibility as well as better aerodynamics and a significantly larger luggage area.

Although the dashboard used the same layout as the then in-production E-Type and Mark 2 and employed the usual quality touches of veneer and leather, this was the most cavernous interior of any Jaguar model before or since. The extra body width and length allowed significant legroom in the rear compartment, without any loss of space in the front or headroom.

At launch in 1961, the Mark X was well received and quite a culture shock for traditionalists. Clearly aimed at the North American market, the car's initial reaction from British buyers was not as good.

model before or after, these were quite normal in North America and a nod to the US market at which the Mark X was clearly aimed. The Mark X was Jaguar's first postwar saloon design without rear-wheel spats.

The whole monocoque was new with every single internal and external panel specifically designed for the car. The longest and widest saloon Jaguar had ever produced, it was also the lowest in height.

The basic structure consisted of two enormous box section sills, the floor (strengthened with large transverse boxes under the front seating area) incorporating the transmission tunnel, bulkhead, and rear seat pan. Two members extended forward from the base of the front bulkhead section to carry the

engine and the front suspension beam braced with diagonal pressings. The rear seat pan consisted of two box-section transverse members. A pressing ran across the car at the bottom of the seat pan, with boxed members underneath to carry the rear suspension assembly. Two further box sections ran rearward to form the sides of the luggage area. The door pillars and their hinges were formed from box sections to carry the considerable weight of the massive doors. The finished structure was immensely strong and rigid—so much so that the roof panel didn't contribute anything to the overall strength of the body.

The Mark X interior was not only luxurious but cavernous, with more usable space than Jaguar had achieved before. Leather upholstery, walnut

veneer (as normal), and a fully equipped dashboard plus a high level of equipment ensured this was a prestige car by any standard.

The rear bench seat easily took three abreast with considerable legroom. The back of the front seats still incorporated veneered occasional tables, now with vanity mirrors and twin ashtrays. The rear passengers had a separate control for the heating/ventilation system.

At the front, separate reclining seats formed a bench-line style, across the significant transmission tunnel covered in leather. The dashboard was of an entirely new design with the main instrumentation in front of the driver, auxiliary gauges and controls on a center panel, and accompanying matching switchgear. This type of dashboard would also feature on all new Jaguar models from 1959 through to the late 1960s. There was a full-width parcel shelf under the dashboard and the vertical section of the center console formed the area for the radio, speakers, ashtray, and heating controls.

The heating system is worth special mention as it too was entirely new with a much larger heater matrix. Two fans boosted the air coming into the car, individually adjustable on both sides of the front passenger area, with further pipes running down the inside of the transmission tunnel to the rear compartment. Just three buttons on the console—labeled Off, Air, and Heat—were controlled by vacuum pressure from the engine manifold.

The steering wheel was of a new design that would not be used on other Jaguars until late in 1964, the column, binnacle, and controls, also new, would be featured on all other new Jaguar saloons from 1959 until the late 1960s. You could also specify your Mark X with electric windows, in which case a veneered panel was set into the center console for the controls.

Mechanically the Mark X was far advanced from the Mark IX and carried over a lot of the features first seen in the E-Type sports car released earlier that year (see Chapter 7). The Mark X was therefore the first saloon fitted with the straight-port cylinder head; indeed the engine was virtually the same as that fitted to the E-Type, down to the triple 2-inch SU carburetors. Minor changes had to be made for the saloon installation, including a new air cleaner/manifold arrangement and engine oil sump.

The engine was available with the usual four-speed manual gearbox, with or without overdrive, or the BorgWarner Model DG three-speed automatic transmission with Jaguar's Intermediate Speed Hold feature. The Mark X also featured the new and sophisticated independent

US COMPARISON

America's love of British cars started during World War II when many GIs stationed in the UK ran the pretty little MG sports cars. They liked them so much that upon their return home they purchased MGs to take with them. Thereafter sales of British sports car blossomed in the US.

Americans also liked British luxury saloons for their drivability and luxurious build, not to mention their street cred—hence Jaguar's success early on with their Mark VII saloon.

The Mark X took eagerly awaiting Americans by surprise . . . it was too similar to their early-1950s Hudson Commodore in both styling and spec:

MODEL	1949 Hudson Commodore	1961 Jaguar Mark X
LENGTH	207.52 in. (5,271 mm)	202.52 in. (5,144 mm)
WIDTH	77.01 in. (1,956 mm)	76.50 in. (1,943 mm)
WHEELBASE	124.02 in. (3,150 mm)	120.00 in. (3,048 mm)
WEIGHT	3,650.9 lbs. (1,656 kg)	3,948.5 lbs. (1,791 kg)

The 3.8-liter Mark X engine with triple 2-inch SU carburetors and the unique Kelsey-Hayes bellows-operated brake servo system (center foreground).

rear suspension cage from the E-Type (again, see Chapter 7).

The front suspension was a development of that first introduced on the Mark 2 in 1959, the full details of which appear in that later chapter. Although using that new design with top and bottom wishbones and a larger coil spring operating on the lower, the Mark X had a single cross beam instead of a subframe.

Because of the small wheel diameter, the braking system was smaller than one might have expected for such a large car. The Dunlop system used 10-inch discs with power assistance from the American-designed Kelsey-Hayes bellows system (also first seen on the E-Type).

Steering followed normal Jaguar practice at the time with a Burman box and power assistance.

Preproduction testing for the most complex saloon Jaguar had produced was more intensive to ensure that when it came to launch, it would fulfill its role admirably.

After the successful E-Type launch in Geneva earlier in the year, the Mark X was launched at the British Motor Show in October 1961 and met with some apprehension because it was so different than any other Jaguar—and any other British car—at the time. The problem for the home market was that the car was so large. Most domestic driveways and garages were not wide enough to accommodate the Mark X, and as for city-center car parks . . . ! Clearly aimed at the North American market, it nevertheless was met with some caution there too. For many it was too "un-British" and considered more like a period Hudson than a traditional British saloon.

Despite concerns, most drivers liked the car for its physical presence paired to performance and handling not usually found in such large cars.

1962–1963: Early Development

As with any new model, Mark X modifications took place throughout production. Following the same changes to other Jaguar models, the engine was modified in 1962 with a revised crankshaft oil seal, drilled camshafts, and improved exhaust valves.

In 1963, the suspension got gas-filled dampers, larger cylinders were fitted in the braking system, the rear suspension mountings were altered, and a new power-steering pump was used. The same year, the automatic transmission unit was revised to improve gear changes and Dunlop radial tires became standard. At the end of 1963, a radiator with a separate header tank area improved cooling in hotter climates.

For the first time, a heated rear screen could be ordered, and the steering wheel center button now operated the horns in addition to the horn ring.

The Mark X launch was cautious and the first full year of production, 1962, amounted to 4,312 cars—remarkably close to the first full year of Mark VII production in the early 1950s. Another 6,572 came off the line in 1963, exceeding the equivalent Mark VII figures.

1964–1965: More Capacity, More Refinement

Increasing the XK engine's capacity to 4.2 liters in October 1964 (see Chapter 7 for more details) brought other changes to the Mark X. The cooling system, for example, with a tube and fin radiator, included a viscous coupling cooling fan that "slipped" at higher speeds to reduce the power it took out of the engine, and a new water pump.

An alternator replaced the dynamo for better charging performance, but necessitated a new power-steering pump, while a Lucas pre-engaged starter motor

POLITICS!

Jaguar purchased the British Daimler company in 1960 and inherited a new 4.5-liter V-8 designed by Edward Turner, a famous motorcycle engineer, for the Majestic Major. Daimler's V-8 had some advantages. It was lighter than Jaguar's 3.8-liter six, produced more torque, was marginally more economical, and it was already in production. The V-8 sat easily in a Mark X bodyshell and produced a significantly higher top speed in tests.

Nevertheless, the V-8 Mark X never went into production. One reason was that it was not geared up for mass production, which would have been a substantial cost. More importantly, the decision was political: the most powerful Jaguar saloon ever built couldn't be powered by a Daimler engine!

improved cold starting. The power steering was a new Marles Varamatic system from Bendix Corporation in the US, adapted by the British Adwest company for Jaguar. As its name implies, it varied the ratio according to the turn of the wheel.

An even bigger change was a new conventional servo for the braking system, operated from the pedal. Although the brake discs remained the same, 4.2-liter cars were fitted with Dunlop Mark III calipers with a larger pad area. Mud shields were also fitted to the front discs.

Although manual transmission Mark Xs were never that popular, Jaguar opted to fit their new in-house four-speed

(ABOVE) Later 4.2-liter Mark X interior with revised heater controls, under-dash pull-out occasional tray, and in this case, power-operated window controls on the center console.

(LEFT) The later 4.2-liter engine in the Mark X with a conventional brake servo system, alternator, and other changes, not least the new Varamatic power-steering system and smoother BorgWarner automatic transmission.

MARK X LIMOUSINE

Jaguar produced a small number of limousine versions of the 4.2-liter Mark X. It was a relatively easy conversion at the factory with no exterior or mechanical changes. Internally, the front seatbacks were replaced with a bench mounted with a sliding glass partition. The B/C post paneling was altered to accommodate the partition, as was the roof-support area and headliner. The veneered occasional tables and cabinet for the rear compartment were also modified with space for a separate radio.

Although the two front seat squabs were still separate, they were no longer adjustable. Front door trims were slightly altered, no longer having upholstered door pulls/armrests.

The limousine option was also available in the later 420G form. A mere 16 4.2-liter Mark Xs and 24 420Gs were produced as limousines.

all-synchromesh unit (see Chapter 7). The automatic transmission was also changed to the BorgWarner Model 8 providing a D2 position for second gear starts. It was a much smoother unit in operation.

Internally, automatic transmission cars featured a new control lever on the steering column that was longer and had a knob at the end. The legend in the column binnacle window also now showed the D2 gear position. The dashboard now had a neat pull-out occasional table under the center section, but more significant were changes to the heating system, with separate heat and air mixture controls for each side of the car, operated from levers set into the front of the parcel shelf.

The "4.2" badge on the boot lid was the only way to discern a 4.2-liter Mark X from a 3.8-liter Mark X, externally.

For the 4.2-liter model, Jaguar worked with Delaney Galley to produce a suitable air-conditioning system. Fitted in the boot with extractor vents on the rear parcel shelf and controls on a panel above the dashboard heater controls, it was expensive and never a popular option.

1966–1970: The Flagship Saloon Era Comes to an End

By 1966, Jaguar had expanded the saloon model range considerably (see Chapter 6). All saloons came in for some changes, including renaming, and the Mark X was now rebadged the 420G.

After a total Mark X production figure of 18,633, some subtle styling changes helped identify the "new" model. Externally, the cars received a new radiator grille with a thicker center spar, side repeater indicator lenses fitted to the front wings, and for single-color bodies, a full-length swage-line chrome trim from the side repeater lights to the rear wings. Jaguar's new-style hubcaps

The revised 420G model now available with a choice of duotone or single-tone schemes. Note the revised radiator grille and "new" side repeater indicator lenses. The gray car in the background is also a 420G, in this case a single color, examples of which featured swage line chrome trim as seen here.

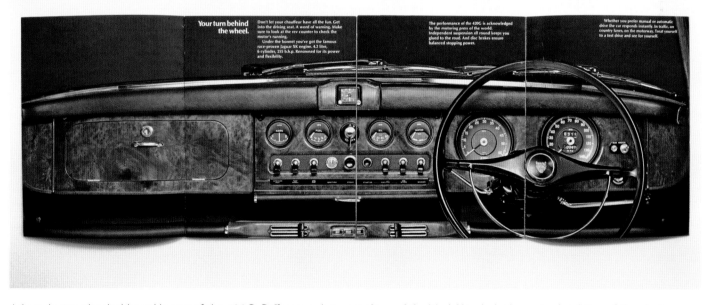

Ad art depicts the dashboard layout of the 420G. Differences between this and the Mark X include the revised sighting of the clock, a modicum of padding on the top dash rail, and amended instrumentation.

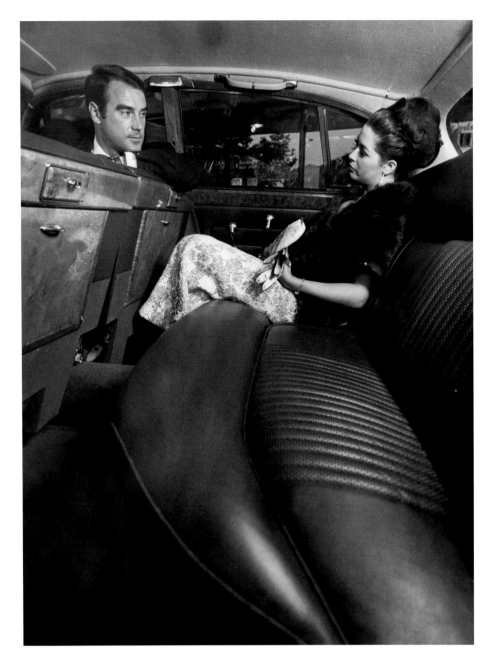

were used, and the "420G" insignia appeared on the boot lid.

Internally, it was much as before except for a new dashboard top rail with black Ambla padded section to the left and right and a square clock fitted in the center. A new rev counter replaced the previous rev counter with clock. The black finish continued to the parcel shelf roll edge and even to the pullout tray before the dash area. The pleated leather seating incorporated aerated panels, a fashionable feature at the time.

The 420G got ribbed cam covers (following the same for all other Jaguar XK engines at the time) and a separate radiator heater tank mounted on the inner wing.

The only marked change came in 1968 when the front engine mountings were transferred from the bodyshell to the front suspension crossmember beam.

The first year of full 420G production, 1967, saw 1,640 leave the assembly line, significantly down on Mark X figures. The trend continued with a mere 600 produced in 1970, although the 420G had stayed in production well beyond the introduction of its successor, the XJ6.

(ABOVE, LEFT) One feature of the 420G was aerated leather seating.

(LEFT) The 420G engine bay with ribbed cam covers and revised header tank arrangement for the cooling system.

THE DAIMLER CONNECTION

The Daimler DS420 limousine, a direct derivative of the Mark X/420G with styling by Sir William Lyons himself.

Jaguar pledged to continue the Daimler brand after its takeover of the company in 1960. The existing Daimler range was quite aged and despite an initial move to produce a limousine version of the Majestic Major, only 867 were made.

Jaguar, then merged with the British Motor Corporation (BMC), had access to the Vanden Plas coachbuilding business in London and needed a new limousine for the specialist and diverse market. William Lyons himself came up with the design for a new car to carry the Daimler name, but substantially underpinned by Jaguar components, the XK engine and the floor pan, inner panels, and many other components from the Mark X/420G.

Initially produced at the Vanden Plas works, it later moved to a special limousine shop at the Jaguar factory. The DS420 was successful in its marketplace. Production ran from 1968 until 1992, with a total of 5,000 produced.

SPECIFICATIONS

MODEL	Jaguar Mark X 3.8	Jaguar Mark X 4.2	Jaguar 420G
ENGINE SIZE	3,781cc	4,235cc	4,235cc
CARBURETION	3 x SU	3 x SU	3 x SU
MAXIMUM BHP	265	265	265
MAXIMUM TORQUE	260@4,000	283@4,000	283@4,000
GEARBOX	4-speed	4-speed	4-speed
AUTOMATIC	BW 3-speed	BW 3-speed	BW 3-speed
0 TO 60 MPH	10.8 sec.	10.4 sec.	10.4 sec.
STANDING ¼ MILE	18.4 sec.	17.4 sec.	17.4 sec.
TOP SPEED	120 mph	122 mph	122 mph
AVERAGE FUEL CONSUMPTION	14 mpg	16 mpg	16 mpg

CHAPTER FIVE

MASS-MARKET COMPACTS

Mark 1, Mark 2, S-Type, and 420

By the mid-1950s, Jaguar had cemented their place in automotive history. The XK sports cars were virtually unequaled in terms of performance and price, and the flagship saloons so well illustrated the slogan "Grace . . . Space . . . and Pace."

William Lyons was still not satisfied and realized the potential to produce a further model aimed at expanding his market sector and using many existing components, thus bringing down production costs for all cars and increasing profitability.

There were few small, luxury saloons around, particularly those offering a standard of performance, handling, and comfort that Jaguar did with the Mark VII. The company had some early success with the 1.5-liter SS Jaguar in the 1930s and 1940s, so as early as 1952 Lyons started work on a new car to meet his criteria. It was a success in sales and in competition, leading to a raft of other models based on it. Variations on the theme continued in production until 1969.

1955–1956: Small Can Be Beautiful

This was an ambitious project for what was still a very small company that already had difficulties in coping with demand for its cars. Given industry developments, Lyons felt in monocoque (chassis-less) construction was the way forward for the new car. It had to be smaller, more economical, and easier to build, and monocoque design seemed the natural progression, with the experience gained ultimately benefiting other Jaguar models. It also had to accommodate the XK engine and other existing mechanical aspects.

The bodyshell comprised two main channel sections running from the very

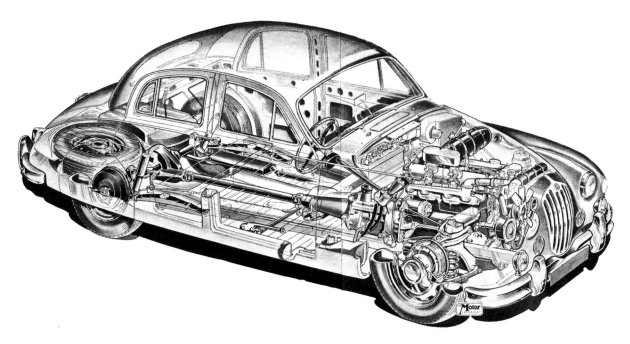

(**ABOVE**) The clean uncluttered lines of the 2.4 with full spats over the rear wheels. Hubcaps were standard fitment but not rimbellisher wheel trim.

(**LEFT**) The 2.4-liter saloon was a major advancement for Jaguar, their first car to feature monocoque construction and offering traditional Jaguar styling with good performance in a smaller package.

The original 2.4-liter saloon with cast radiator grille and full rear-wheel spats.

front of the car to the rear-wheel arches and welded to the floor to form box sections. The floor was ribbed for extra strength and transverse members joined the longitudinal members, with the whole structure tied together by the front bulkhead and rear seat pan. Two further box sections ran diagonally up each side of the engine bay to transfer stress via the A-post pillars to the roof area. All exterior and interior paneling was welded to the main structure with the outer sills contributing to overall strength.

The styling took cues from the existing models with an all-enveloping design, the front end particularly reminiscent of the XK140. Indeed, features from the launch of the XK140 found their way into the new small saloon, like the bumper bars, headlights, rear lights, and although subtly different, even the cast radiator grille. Unique oval sidelights were fitted to the base of the front wings.

The bonnet as usual featured a central chrome strip and Jaguar leaping mascot, although not for some overseas markets. The body had a gentle curvature at waist height, highlighted by chrome strips running from the bonnet through the door lines right down the rear wings to the bumper bar. Full spats covered the rear wheels.

New features included doors incorporating the window frames like the pre–Mark V saloons—a throwback but apparently done to give rigidity to the bodyshell. The door handles were also unique to this model.

At the rear, the new car was a scaled-down version of the Mark VII in many ways but with a more tapered line from the sides.

Mechanically, a smaller car could use a smaller engine. Although a four-cylinder version of the XK engine was considered, this was too costly to produce and not smooth enough. Keeping the six-cylinder XK, but in a

The smaller 2,483cc XK engine developed for the 2.4-liter saloon, with twin Solex carburetors, a rare move away from their traditional SU type.

The 2.4-liter launch at the British Motor Show in London, featured alongside the XK140 sports, and in the background the Mark VIIM saloon.

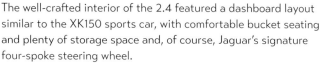

The well-crafted interior of the 2.4 featured a dashboard layout similar to the XK150 sports car, with comfortable bucket seating and plenty of storage space and, of course, Jaguar's signature four-spoke steering wheel.

For a small luxury car, rear compartment space was good. Note the full door frames incorporating the window surrounds, finished off with wood veneer.

different configuration, could keep costs down and be much easier to get into production.

To achieve the final configuration of 2,483cc, the same bore from the 3.4-liter unit was used but with a shorter stroke of 76.5 millimeters. Reduction of engine block depth resulted in considerable weight savings. A financial benefit came in using the same bottom-end internals as the bigger engine. The twin cam cylinder head was also retained but with different camshafts. Induction was new with twin Zenith Solex downdraft B32 carburetors with a revised inlet manifold.

In this form the new 2.4-liter engine developed 112 brake horsepower at 5,750 rpm, with a maximum torque of 140 pounds-feet at 2,000 rpm. A Borg and Beck single-plate clutch drove the power to the usual four-speed Moss-derived gearbox, with overdrive available at extra cost.

Naturally the engine was a tight fit; the taper of the front wings and

bonnet caused particular concerns. With the battery mounted on the bulkhead along with the heater box there was just enough room in the center to accommodate the back of the engine! The air cleaner had to be adapted by using a trunking across the top of the engine to exit through the wing to an oil bath arrangement.

The boot area was inevitably smaller than the Mark VII's. Retaining the rear hinged boot lid, a simpler lock was employed, and the interior of the lid was unlined. The fuel tank was mounted underneath the boot floor, shaped to fit around the spare wheel well, the latter accessed by removing a metal panel in the boot floor. A new toolkit sat inside the inner face of the spare wheel and a tailored Hardura mat covered the boot floor, leaving the rear wheel arches and their damper locations visible. Board panels sealed off the inside of the rear wings, the left-hand side having the fuel pump and filler pipe.

Conventional coil springs formed the basis of the front suspension with two rear-inclined unequal-length wishbones, the spring held in the lower wishbone by a hollow steel pillar on which the top wishbone pivoted. Upper and lower ball joints accommodated wheel movement and steering. Shock absorbers were fitted inside the top of the suspension pillar running through the springs to the bottom wishbone, which was linked to its opposite number by an antiroll bar.

The whole of the front suspension was mounted on a separate subframe secured to the body by rubber mountings to avoid the transference of noise and vibration.

Steering was by the usual Burman recirculating box, also mounted on that subframe. It was insulated from the steering wheel by two universal joints; rubber bushes were used for the inner wishbone mountings, all to avoid noise and vibration passing through the monocoque.

SETTING THE STANDARD

The 2.4 was marketed as the Special Equipment model, and initially Jaguar also offered a Standard version without such features as a bonnet mascot, auxiliary lighting, heater, screen washers, rear seat armrest, cigarette lighter, or rev counter. At a mere £29 less it was never going to be a success, so in truth no known examples of the Standard were actually produced.

At the rear, the axle was cantilevered from mountings located in the main monocoque structure. Further location was provided via two trailing arms running from the seat pan to brackets above the axle, plus an adjustable Panhard rod. Conventional leaf springs were fitted upside down with the trailing ends attached via rubber bushings to extensions in the axle casing. They anchored to the car inside the rear of the chassis rails. It used Girling telescopic dampers and Salisbury differential.

Lockheed drum brakes were fitted to the new car, with self-adjusting front shoes. For the first time on a production Jaguar, 15-inch wheels and tires were fitted.

A unique feature of the new car was its narrow rear track (4 feet 2¼ inches) compared to the front (4 feet 6¾ inches). Apparently, this feature was dictated by the body taper at the rear end, which also created handling issues. Although the car looked rather odd when the rear wheel spats were removed, with the spats on, one could hardly tell the difference.

Based on the new engine size, Jaguar decided to call the model the 2.4. The package met all of Lyons's requirements for the small car—indeed, it was 16 inches shorter than the Mark VII and 8 hundredweight (800 pounds) lighter.

Despite the cars being smaller and more economical, Jaguar didn't let up on the equipment levels. Internally, it retained wood veneer—and got more of it, with the door frames surrounded in wood—plus leather upholstery and full instrumentation.

Small individual bucket seats featured in the front with the usual bench in the rear, and leg and headroom were remarkable for such a small car, although visibility was somewhat hindered by the small rear screen and reduced side windows. The dashboard layout was new, with a lidded glove box on the passenger side and an open cubby on the driver's side. The smaller Bluemels steering wheel retained the Jaguar four-spoke design, and a unique feature was the hidden ashtray within the veneered fillet just below the ignition key switch.

Jaguar launched the 2.4 at the 1955 British Motor Show to some acclaim, although it wasn't so well received in the US, where smaller cars were not very popular, and performance was more important than fuel economy. Although the 2.4 was a good performer by the standards of the day, it struggled to reach a top speed of 100 miles per hour and acceleration was not in the league of larger-engined cars. Nevertheless, sales got underway very quickly and in huge numbers.

There were inevitable teething troubles. Early in 1956 production, the Panhard rods were found to be breaking, so a strengthening plate was added and the end of the rods made adjustable to achieve correct tension. Stiffer dampers were also required. Various minor changes affected the new engine with a steel sump replacing alloy, and the installation of a metalastic crankshaft damper.

Sales were buoyant in that first year of full production: a total of 8,061 cars were delivered, nearly twice as many as all the other Jaguar models sold at the time. To address concerns about the lack of performance, Jaguar prepared a special booklet identifying three stages of tuning modifications for both domestic and competitive driving. Stage 1 increased power to 119 brake horsepower, Stage 2 to 131 brake horsepower, and Stage 3 to 150 brake horsepower.

1957: Power to the People

As a company known for performance, Jaguar had to heed concerns that the 2.4 was underpowered. Problem solvers could not help but consider the 3.4-liter XK engine as a potential answer. This swap transpired in February 1957 with an assortment of modifications.

The transplanted engine had the B-Type cylinder head with the twin SU HD6 carburetors, but with a new inlet manifold and air cleaner. Mounts had to be strengthened and a larger radiator was installed. A larger 10-inch clutch was fitted along with a new V-mount under the gearbox.

To improve cooling efficiency, a new wider radiator grille was fitted with narrower slats (incorporating a 3.4-liter badge), which also necessitated a modified front wing design. This type of grille was also standardized for the 2.4 from September of that year.

To aid cooling to the rear brakes, the full spats were abandoned, replaced by a cut-away type. A "3.4" badge appeared in the boot lid and the car was equipped with a twin exhaust pipe system.

Jaguar offered the 3.4 with the BorgWarner Model DG automatic transmission, complete with Jaguar's

Intermediate Speed Hold feature, as an option for that model. A central gear selector worked for drivers on the right or left. A new split-bench front seat was fitted to automatic transmission cars.

The 3.4 proved to be an excellent performer, with a 0- to 60-miles-per-hour acceleration time of 9.1 seconds compared to the 2.4's 14.4 seconds. The car was well received, even if Jaguar couldn't meet early demand due to a fire at the factory in the same month the car was launched. Right from the start, however, there were concerns, particularly from the motoring press that with such performance, the drum brakes were totally inadequate. Jaguar had already been working on disc brakes for the XK150 sports car, but now had to bring the installation forward for the 3.4.

Disc brakes became an option shortly afterwards. Twelve-inch discs were fitted all round but with the earlier Dunlop round pads, which required dismantling the cylinders to change them. Servo assistance was also fitted.

(ABOVE, LEFT) The automatic-transmission selector lever was placed here below the center of the dashboard, for ease of access in both left- and right-hand drive form.

(CENTER & BOTTOM, LEFT) Showing the styling differences between the Mark 1 in its later 3.4-liter form and the replacement car, the Mark 2 in 1959.

(OPPOSITE, TOP) US promotional picture of the 3.4-liter saloon when it was announced in 1957.

(OPPOSITE, BOTTOM) Comparison frontal treatment for the first 2.4-liter model (left) and the revised radiator grille treatment, initially for the 3.4-liter and then fitted to 2.4s as well.

Squadron "A" Armory, N.Y.

1958–1959: Enhancing the Product

Jaguar was learning a lot about the small, sporting saloon market. The 2.4 concept had been expanded with the 3.4, and now an automatic transmission was available. Wire wheels became an option in a range of finishes, initially 60 spoke, later 72. The "new" disc brakes were available as an after-market kit to fit earlier drum braked cars, from February with a larger brake servo. In January 1959, the braking system was further changed for the Dunlop square "quick-change" pad type. The Dunlop tire design was changed to improve wet weather grip and a Thornton Powr-Lok differential became an option.

Later changes included a 12-bladed cooling fan, and a short-stud cylinder head, common to other Jaguar models at the time.

The 3.4 outsold the 2.4 in each subsequent year of production. Final totals of 19,705 2.4s and 17,280 3.4s made them Jaguar's most successful models up to that time, despite being in production for such a short period of time.

The high-performance 3.4-liter with wire wheels, necessitating the replacement of the full rear spats with cutaways, later adopted for the 2.4 as well.

1959: Let There Be Light (and More Performance)

After three years, Jaguar decided that it wanted more than a facelift of the 2.4/3.4 saloons. The Mark 2 was smarter, with a cleaner and lighter interior and improved handling. There was also an extra version with more power.

At first glance from the front, the most prominent change was a new radiator grille with thick center bar containing a mounting for the engine-size badge. The lighting had also changed. Rear mounted auxiliary lights of a new Lucas design were inset where the horn grilles used to be. Some markets, including the US, did not receive the auxiliary lights and instead got chromed mock horn grilles. Sidelights sat in pods on top of the wings but now with separate indicator lights at a lower level near the bumper bar—the same as featured on later XKs and Mark VIIM onwards.

The glass area was enlarged with a new windscreen. Gone were the steel-framed doors of the previous model, replaced with the traditional chromed brass window frames, with a larger glass area. Similarly the rear screen was enlarged. The rain gutters in the roof, previously an integral part of the metal and painted, were now covered with a chrome finisher.

At the rear, the badging on the boot not only included the engine size but also the discrete "Mk 2" badging. The bumper bar itself may have looked like the earlier cars but the curvature was subtly different as was the shaping of the rear wings, necessary to accommodate a wider rear track. The other change was in the rear light units. They were much larger to accommodate separate indicator lenses and on deeper chromed plinths.

Internally, there was an all new dashboard design, a style that Jaguar would use on all models through the 1960s, including the sports cars. Very businesslike, still in wood veneer, it featured the main speedometer/rev counter in front of the driver, with auxiliary gauges mounted on a separate black panel centered in the dashboard. On the passenger side there was a lidded glove box. An upholstered center console ran down from the dash area between the front

The Mark 2 in its original form with steel wheels.

bucket seats into which were fitted the heater slider controls, radio and speaker, plus an upholstered ashtray.

There was also a new steering wheel design with a single center spar and incorporating a half horn ring. As with all other Jaguars, the steering wheel was adjustable for reach and the new wheel incorporated a column binnacle with a window revealing an illuminated legend for either overdrive or automatic transmission cars. The automatic transmission or overdrive was operated by a stalk on one side; on the other side was a matching stalk for the indicators, which could be pulled to flash the headlights.

Upholstery was still leather faced with new, larger front bucket seats. Front door panels incorporated elasticated map pockets, while rear doors held solid pockets and incorporated ashtrays.

For the rear compartment, occasional pull-down tables were incorporated into the back of the front seats, upholstered on the outside but with a wood veneered panel on the inside. Rear bench seating was subtly different from the previous model and the interior roof lighting was also of different design. The whole interior effused modernity and lightness. No changes were made to the boot interior.

The heating system was improved with a higher rated output and there was now a separate heater duct piped through the center console to the rear compartment.

Mechanically, the Mark 2 was a much-improved car. The axle casing had been extended on both sides to provide an extra 3½ inches in the rear track. This not only improved the appearance of the car but also helped the handling. Further, angling the front wishbones down lifted the car's roll center, also improving the handling.

Most of the power train was unchanged from the previous model, except that the 3.8-liter engine was available in a top-of-the-range, high-performance version of the Mark 2. With suitable changes to accommodate the engine in cramped confines, including a pancake air cleaner on top, this model produced 220 brake horsepower and a genuine top speed of over 120 miles per hour, making it the fastest production saloon in the world at the time.

The comprehensive dashboard layout of the Mark 2, which would be used for all future Jaguar models up to the 1970s. The main rev counter/speedometer in front of the driver, all auxiliary instrumentation and switchgear on the center panel, with a center console for heater and radio controls.

Examples of the extremes in engine performance for the Mark 2: on the left, the 2.4-liter model, which sat lower in the engine bay and featured twin Solex carburetors with an oil bath air cleaner; and on the right, the 3.8-liter model with twin SU carburetors and conventional pancake air filter arrangement.

(**OPPOSITE, TOP**) The Mark 2 as launched for some overseas countries with auxiliary lighting replaced by mock horn grilles.

(**OPPOSITE, BOTTOM**) The launch of the Mark 2 at the British Motor Show in London, here seen alongside the Mark 1, which was still listed as a new car at the time, probably to run stocks down.

(**RIGHT**) Color-coordinated badging on the Mark 1 and Mark 2 grilles to identify the engine size in the car.

The ultimate variant of the Mark 2, the 3.8-liter model with wire wheels—when released, the fastest production saloon car in the world.

Jaguar showed the world the Mark 2 at the British Motor Show in October 1959, where it was very well received. It proved highly competitive in the marketplace and immediately became the car to own for anyone who wanted prestige with performance in a relatively small package.

Despite its late launch in the year, Jaguar was well geared up for production. All three engine sizes were available immediately and no fewer than 2,532 were produced in those first few months.

1960–1962: Exceeding All Expectations

The first couple years of Mark 2 production exceeded Jaguar's expectations. It became the most popular and profitable model they had produced at the time, with sales over 40,000 in the first two years!

There were on-going improvements. A stronger anti-roll bar, stiffer dampers, forged wishbones (instead of pressed) and a self-adjusting handbrake all came into production. Engine oil lubrication was improved, a higher output dynamo and sealed beam headlights were introduced for certain markets.

Internally, an organ type acceleration pedal replaced the earlier type, the twin stalks on the steering column swapped positions to match internationally accepted standards, and the oil pressure gauge calibration changed to make it more accurate to read. The sun visors were changed, the black center dash panel was now fitted with a Rexine fabric finish, the rearview mirror was revised, and strengtheners were added to the window frames. In addition, the front seat backs were reshaped at the bottom to allow more legroom for rear seat passengers.

THE GOLD STANDARD

So important was the Mark 2 to Jaguar's plans to expand sales that the company clad the display model at the 1960 New York Motor Show in a metal sure to go unmissed: gold.

The centerpiece of Jaguar's stand at the show was a 3.8-liter Mark 2, extensively gold plated, with an estimated value of $25,000—when a production car would have cost just $5,000!

Every part that could be gold plated was done, from bumper bars to wire wheels, from wiper blades to the ignition key. Even the accompaning model wore a 24-carat gold braid dress and a tiara containing 1,000 diamonds.

The car was "normalized" after the show and sold off as a conventional Mark 2!

The Mark 1 and Mark 2 were very competitive in the late 1950s and early 1960s in both racing and rallying competition.

(ABOVE) Trim differences between the Mark 2 (in tan) and the Daimler V-8 (in gray).

(TOP) The Daimler 2.5-liter V-8 model was based on the Jaguar Mark 2, but with a 2.5-liter Daimler V-8 engine and traditional Daimler trim—a car that created a new and different market for the company.

(BOTTOM) The V-8 engine was a neat fit in the Mark 2 bodyshell and saved some weight, producing good performance compared to the Jaguar 2.4-liter model.

A MARK 2 ESTATE?

Although Jaguar never produced an estate car before the X-Type in 2004, a single Mark 2 was privately commissioned in the 1960s. Named the County and built by Jones Brothers Ltd., the estate retained a four-door arrangement and had a conventional opening tailgate.

Shortly after it was produced, the estate was purchased by Jaguar Cars but never progressed as a production idea. Instead, it was retained for some years as a support vehicle for Jaguar's racing exploits. Eventually sold, the car is still around today in private ownership after going through a complete restoration.

(RIGHT) The scripted "D" to form a bonnet adornment on the Daimler.

1962: A Daimler Derivative

The Mark 2 was thriving about the time that Jaguar had purchased the Daimler Motor Company, the UK's oldest car maker. Lyons had vowed not only to continue production but to revitalize the range.

Daimler used a 2.5-liter V-8 engine in their low production SP250 sports car. It was a well-designed power plant, boasting 140 brake horsepower at 5,800 rpm. When Jaguar tested this engine in its 3.4-liter bodyshell, the results were impressive—suggesting both a new car for Daimler and an extra model in Jaguar's range.

Because the V-8 was lighter, Mark 2 spring and damper rates were altered, and steering became lighter too. In performance terms, the V-8-powered car was no match for the larger engine Jaguars, but it could exceed performance of the 2.4 liter. Initially, the collaboration car was fitted with only a BorgWarner Model 35 automatic transmission.

To "Daimlerize" the car, it was given a traditional fluted radiator grille and badging, matched by similar treatment at the rear. Off the bonnet came the Jaguar leaper, replaced with a stylized "D" and similar treatment was given to the hubcaps and other areas of the car. At the rear, the Daimler sported twin exhaust pipes with one existing at each side instead of together.

Internally, although the basic design from the Mark 2 was retained, different styling was given to the door trims and seating, with the front forming a split-bench arrangement. Gone was the Jaguar's center console, replaced by a vertical veneered and trimmed section suspended from the center of the dashboard and containing the radio and heater controls.

The Daimler 2.5-liter V-8 expanded Jaguar's customer base, drawing buyers loyal to the other badge.

'Superb–there's not a car in its class to touch it...'

SUNDAY EXPRESS

Outwardly the Jaguar 'S' type saloon is a car of compact dimensions—but when you step inside you will discover spaciousness quite remarkable in a medium sized car. Generous head and leg room and lavish furnishings conspire to offer expensive 'big car' comfort for driver and passengers alike—and there is a luggage boot of no less than 19 cubic feet to match. But, above all, the 'S' model, powered by the 3.4 or 3.8 litre twin overhead camshaft XK engine, 5 times winner of Le Mans, includes the very latest advancements in Jaguar engineering and is available with automatic transmission or new 4-speed all-synchromesh gearbox and overdrive. Fully independent suspension, self-adjusting disc brakes on all 4 wheels and self-adjusting handbrake, driver operated variable

interior heating for front and rear compartments, reclining front seats, twin petrol tanks and a host of other features provide typical Jaguar travel . . . that special kind of motoring which no other car in the world can offer.

3.4 & 3.8 'S' MODELS

The Jaguar range also includes the 4.2 Mark Ten Saloon, the 2.4, 3.4 and 3.8 Mark 2 models and the 4.2 'E' type models

LONDON SHOWROOMS: 88 PICCADILLY W.1

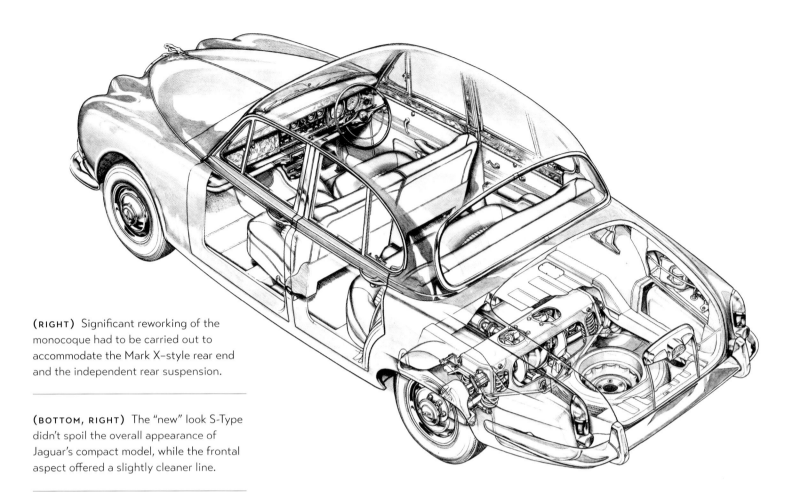

(RIGHT) Significant reworking of the monocoque had to be carried out to accommodate the Mark X–style rear end and the independent rear suspension.

(BOTTOM, RIGHT) The "new" look S-Type didn't spoil the overall appearance of Jaguar's compact model, while the frontal aspect offered a slightly cleaner line.

(OPPOSITE) The launch of the S-Type in 1963 provided an extra model for those who wanted more space and a better ride than the Mark 2 but didn't need the gargantuan Mark X.

1963: S for an Extra Model

Jaguar's "new" 1960s saloon range was proving very successful. The Mark 2 satisfied demand for a smaller and more sporting model, while the Mark X fitted the bill for those who needed a bigger and more prestigious car. Considering both cars, Lyons concluded that a third model could be produced to fill a gap between the Mark 2 and its big brother.

Incorporating traditional design and features with newer technology and styling, the S-Type was still significantly based on the Mark 2 in many areas. It also borrowed from the Mark X's rear styling to allow extra luggage accommodation, and the E-Type/Mark X's rear suspension to improve ride and handling.

The center body section was like the Mark 2 but with a flattened, slightly expanded roof panel to accommodate a more upright rear screen. The doors were not structurally altered but the hinges and detail fitment were. At the front, the wings were adapted to take a new lighting design. The headlights were extended to create a lip like an eyebrow over them. Auxiliary lights were also set into the wings to create an eyebrow effect. (As with the Mark 2, for certain markets these were left off in favor of a new style of mock horn grille.) The front

wing pod-mounted sidelights had finally disappeared in favor of integrated side/indicator lenses similar to the Mark X. A slimline bumper bar and overrider arrangement also resembled the Mark X, with an indent to accommodate the bottom of the new radiator grille, featuring a bolder surround than the Mark 2s.

The biggest styling change came at the rear with an extended boot area to provide more luggage space. Rear trim included the slimline bumper treatment, revised boot lid and the twin exhaust pipes exiting in the center of the car, also like the Mark X.

The changes rear of the doors in styling and suspension meant a major redesign of the monocoque, as this area of the Mark 2 shell was totally unstressed. To achieve the necessary strength, the box sections that normally contained the rear leaf springs were extended further back over the wheel arches and under the boot floor. The double-skinned boot floor was welded to these box sections.

Other changes at the rear included twin fuel tanks in the rear wings, increasing fuel capacity, and relocation of the under-floor spare wheel compartment to the center.

All this meant that the S-Type could be fitted with the all-independent rear suspension cage from the E-Type/Mark X, a major improvement in ride comfort and handling over the leaf sprung Mark 2.

(ABOVE) A useful comparison in styling between the Mark 2 and S-Type. Note particularly the different rear screen and flatter roofline.

(LEFT) Comparisons of the frontal aspects of the Mark 2 and S-Type.

(LEFT) Overhead views of the Mark 2 and S-Type illustrating the different styling approach.

all-synchromesh gearbox, a year later than it appeared in other models.

1966: Another Addition to the Family

By 1966, Jaguar's medium size saloon sales were falling, while more competitive and modern cars were entering the market. Jaguar's line had been developed in the 1950s, so something entirely new was in order. That process would take time, however, so there was no option other than revamping existing cars in the interim.

In October, Jaguar introduced another new saloon, the 420, with an equivalent Daimler, the Sovereign. Externally, from the bulkhead/windscreen back, the car was pure S-Type in all respects except for the badging on the boot lid. The front was a complete redesign, with new flatter front wings and bonnet, Mark X lighting and bumper bar, rectangular mock horn grilles, and yet another style of radiator grille, slighter taller than the Mark X, with the thick center slat like the 420G. A Mark X's smaller Jaguar leaper adorned a front panel ahead of the bonnet.

Internally, there were many detail changes. The center auxiliary instruments/switch panel, still finished in veneer like the S-Type, was flush fitted to match the passenger side glove box and driver's side instrument panels. There was a full-width dashboard top surround in dense foam covered in black vinyl that continued along the sides and underside of the dashboard its entire width. In the top center of this was a square clock like that used on the 420G. There was a full-width parcel shelf with matching black trim, with the S-Type heater controls set into the black trim of the parcel shelf.

Mechanically, the rest of the car was similar to the Mark 2, although no 2.4-liter version was offered because the extra weight would have affected performance. Mark 2s had always been available with power-assisted steering as an option and this was continued with the S-Type, although improved with lighter gearing.

The S-Type interior was another crossover of Mark 2 and Mark X designs with the general layout unchanged. The dashboard center section was veneered like the Mark X and a similar center console contained the radio and vacuum controls for the heater.

Door panels were unique to the model. Split-bench front seats and rear bench seats were of a new design. Like the Daimler V-8, there were no occasional tables built into the rear of the front seats.

1964–1965: Record Sales

The S-Type proved to be an excellent and successful addition to Jaguar's medium-sized saloon range. In 1964 alone 8,000 were sold, 1,000 more than the Mark 2. The Daimler V-8 had also been a success with nearly 4,000 sold that year. Total Jaguar production for 1964 was nearly 26,000 cars, a record for the company in a single year.

During this period there was the inevitable cross-fertilization of upgrades and fixes across the Jaguar model range. One of the most significant changes took place in September 1965 when the S-Type received a four-speed

(**ABOVE**) From the front screen back, the 420 was entirely S-Type.

(**LEFT**) The introduction of the 420 models in 1966 provided another range of cars to boost the Jaguar range, but also to provide a stop-gap while the next entirely new model (the XJ6) was still under development. The frontal view of the 420 clearly resembles the more modern looks of the Mark X.

(ABOVE) Seating in the 420 was similar to the S-Type, but later aerated upholstery was used.

(LEFT) Yet another interior revision from S-Type to 420. Note the padding completely surrounding the dashboard.

The center console, seating, and door panels were much as the S-Type except for the latter where the wood veneered tops were replaced with cushioned trim, with a flat fillet of wood on top.

The biggest change came with the 4.2-liter version of the XK power unit first seen in the E-Type and Mark X. Where those cars were fitted with triple carburetors, the 420 only had two SU HD8s, a new inlet manifold, and retained the straight-port cylinder head.

Cooling was improved over the S-Type with a new crossflow radiator with a higher capacity water pump. The usual choice of transmission included the four-speed all synchromesh manual gearbox or BorgWarner Model 8 automatic. Power steering came from the same Marles Varamatic system first seen on the Mark X; brakes were by Girling with independent circuits front and rear.

The Daimler Sovereign was a rebadged 420, the only changes being the fluted radiator grille, rear boot lid nacelle, and a scripted "D" replacing the leaper on the bonnet.

For 1966, Jaguar offered seven medium-size saloon variants: three Mark 2s, two S-Types, 2 420s, plus a Daimler 2.5-liter V-8. All these added up to a total of 13,967 car, representing 60 percent of the total company outlet of vehicles. Mark 2 sales were now half those produced by the S-Type, with the 3.8-liter Mark 2 suffering the most.

1967–1969: Rationalization

By the beginning of 1967, the writing was already on the wall for some of the smaller saloons. To keep costs down, Ambla had replaced leather upholstery as standard on Mark 2s. The auxiliary lights on the front of the Mark 2 for most markets were now also deleted as standard equipment, replaced by horn grilles. However, there was an improvement with the Varamatic power steering finally offered on all Mark 2s beginning in July.

Greater changes took place in September 1967 with a rationalization of Jaguar's entire saloon range. The 3.8-liter Mark 2 was discontinued and the 2.4/3.4 were partially facelifted and renamed 240 and 340.

Externally, the renamed cars were identifiable by S-Type style slimline bumper bars. These required new valances front and rear as they were more visible. Badging of course changed and Jaguar throughout all their steel wheeled cars adopted a new style of hubcap with a prominently raised center containing a jaguar head or Daimler "D" emblem.

Both the 240 and 340 continued with Ambla upholstery and lost the occasional tables fitted in the front seat backs.

Mechanically there were bigger changes. The straight-port cylinder head was now fitted to both models and on the 240, twin SU HS6 1¾-inch carburetors replaced the old Solex type, with a conventional paper air filter arrangement

The face lifted 240/340 models with slimline bumpers and reduced trim levels.

340 3.8S

There was never officially a 3.8-liter Mark 2 after the mild facelift to 240/340 form. Sales of the 3.8-liter version had been falling, particularly with the introduction of the S-Type and later the 4.2-liter-engined 420.

Paying customers could order whatever they wanted, however, and certainly a small number ordered their 340s with the 3.8-liter engine. These were fitted with the straight-port cylinder head and ribbed cam covers, as well as the four-speed all synchromesh gearbox, providing a good increase in performance over the Mark 2.

Leather moved to an extra-cost option, while removing the occasional tables in the back of the front seats produced additional savings.

(ABOVE) The later installation of the XK engine in all these cars featured a straight-port cylinder head and ribbed cam covers.

(LEFT) A later V-8–250 Daimler interior, showing the lack of a transmission tunnel, but with a drop-down area from the dashboard to accommodate the heater controls and radio. The padded top roll on the dashboard was also a later feature. This car is one of the rare manual-transmission models.

A later Daimler V-8-250 with slimline bumpers, yet retaining auxiliary lighting and the wheels' rimbellishers.

replacing the old oil bath. The 240 got a twin-pipe exhaust system, too, for a total power increase of 13 brake horsepower.

The changes reduced the price, leaving the 240 a mere £20 or about $50 dearer than the original 2.4-liter saloon in 1956! The 340 was also similarly reduced in price over its Mark 2 equivalent.

During this "updating" period the Daimler 2.5-liter V-8 was revised and renamed the V-8-250. Daimler got the same slimline bumper and hubcap changes but retained the auxiliary lighting. Internally, a black padded roll was

applied to the top dashboard panel. The door panels were finished like the 420 with a top padded roll and single wood fillet on the top. Leather seating, reclining at the front, and a heated rear screen were standard on the V-8-250.

In the engine compartment, twin air filters replaced a single one, and an alternator and power-assisted steering were standard. Now the Daimler could also be specified with the four-speed all-synchromesh manual gearbox as well as the BorgWarner Model 35 automatic.

The S-Types didn't go without changes either, though not as marked

as the Mark 2s. Ambla upholstery and new style hubcaps were standard, the car received the all-synchromesh gearbox, and horn grilles replaced the auxiliary front lights. These changes dropped the price by around £100 or $250 per car.

The revised vehicles were short-lived. By the end of 1968 the 340, 420, and both S-Types were discontinued, followed by the 240, V-8-250, and Daimler Sovereign (420) in 1969. Jaguar produced a grand total of 186,888 compact saloons from 1956 through 1969. It was the end of the smaller Jaguar saloon until 1999.

SPECIFICATIONS

MODEL	Jaguar 2.4 Mk 1	Jaguar 3.4 Mk 1	Jaguar 2.4 Mk 2	Jaguar 3.4 Mk 2	Jaguar 3.8 Mk 2
ENGINE SIZE	2,483cc	3,442cc	2,483cc	3,442cc	3,781cc
CARBURETION	2 x Solex	2 x SU	2 x Solex	2 x SU	2 x SU
MAXIMUM BHP	112@5,750	210@5,500	120@5,750	210@5,500	220@5,500
MAXIMUM TORQUE	140@2,000	216@3,000	144@2,000	216@3,000	240@3,000
GEARBOX	4-speed	4-speed	4-speed	4-speed	4-speed
AUTOMATIC	BW 3-speed	BW 3-speed	BW 3-speed	BW 3-speed	BW 3-speed
0 TO 60 MPH	14.4 sec.	9.1 sec.	17.3 sec.	11.9 sec.	8.5 sec.
STANDING ¼ MILE	24.6 sec.	17.2 sec.	20.8 sec.	19.1 sec.	16.3 sec.
TOP SPEED	98 mph	120 mph	96 mph	119 mph	125 mph
AVERAGE FUEL CONSUMPTION	18.8 mpg	16 mpg	18 mpg	16 mpg	15.7 mpg

MODEL	Daimler V-8	Jaguar 3.4 S-Type	Jaguar 3.8 S-Type
ENGINE SIZE	2,548cc	3,442cc	3,781cc
CARBURETION	2 x SU	2 x SU	2 x SU
MAXIMUM BHP	140@5,800	210@5,500	220@5,500
MAXIMUM TORQUE	155@3,600	216@3,000	240@3,000
GEARBOX	n/a	4-speed	4-speed
AUTOMATIC	BW 3-speed	BW 3-speed	BW 3-speed
0 TO 60 MPH	13.5 sec.	13.9 sec.	10.2 sec.
STANDING ¼ MILE	23 sec.	19.2 sec.	17.1 sec.
TOP SPEED	110 mph	120 mph	121 mph
AVERAGE FUEL CONSUMPTION	17 mpg	15 mpg	15 mpg

SPECIFICATIONS

MODEL	Jaguar 240	Jaguar 340	Jaguar/Daimler 420 Sovereign
ENGINE SIZE	2,483cc	3,442cc	4,235cc
CARBURETION	2 x SU	2 x SU	2 x SU
MAXIMUM BHP	133@5,500	210@5,500	245@5,500
MAXIMUM TORQUE	146@3,700	216@3,000	283@3,750
GEARBOX	4-speed	4-speed	4-speed
AUTOMATIC	BW 3-speed	BW 3-speed	BW 3-speed
0 TO 60 MPH	12.5 sec.	8.8 sec.	9.9 sec.
STANDING ¼ MILE	18.7 sec.	17.2 sec.	16.7 sec.
TOP SPEED	106 mph	124 mph	123 mph
AVERAGE FUEL CONSUMPTION	18.4 mpg	17 mpg	15.7 mpg

CHAPTER SIX

THE ICONIC BRITISH SPORTS CAR

E-Type

The E-Type has become the benchmark by which all other sports cars in the world have been measured. As Enzo Ferrari once said, "It is the best looking car in the world."

Nothing has approached the E-Type in terms of purity of style or world appreciation. Its overall package of performance, practicality, comfort, and good looks plus value for money when new, and investment potential in recent years, shows that it has stood the test of time, a true icon.

1961: A New GT is Born

Jaguar's E-Type was so revolutionary in many ways. At the heart of its success was William Lyons's ability to know when something was right. At its launch at the Geneva Motor Show in March 1961, it left the world speechless for a moment, then unable to keep quiet, creating more rapture than even the XK120 had done in 1948. Although the E-Type retained the XK engine and gearbox, virtually everything else was new, a lot of which had come about through racing experience.

Describing the E-Type, one must start with the ravishing and aerodynamic body, a monocoque of advanced design. The bonnet was made up of a huge center section, flanked by separate wings. The front section led down to an aerodynamic "nose" with a mouth instead of a radiator grille. The bonnet hinged from the front, lifting in one piece to give unobstructed access to the engine and front suspension. A framework of Reynolds 541 square tubing bolted to the bulkhead extended forward to carry the engine and front suspension, with a further frame to carry the radiator and bonnet anchorage/hinging.

A curved single-piece, quite shallow windscreen required no fewer than three windscreen wipers to clear it without rising over the top edge.

The whole of the passenger area from the bulkhead back to the rear quarters was made up of welded steel panels. Large hollow sills started just behind the front wheels, continued under the bulkhead and doors to the rear wing area to mate with a transverse hollow section before the rear suspension. In the floor the propshaft tunnel provided extra strength with a further transverse member running across under the seats.

A well-known image of the original E-Type taken from the launch brochure for the model—an artist's interpretation adapted from a real car photographed in the studio. This roadster displays the fiberglass hardtop that could be ordered from the factory and painted to body color.

Two more longitudinal members ran under the floor from the bulkhead to the rear crossmember. The boot floor, with its two longitudinal box sections either side, joined up with that rear crossmember, all of which took the rear suspension subframe. On to all this the exterior curvaceous paneling was welded to form an exceptionally strong structure for both roadster and fixed head coupe styles.

Stylistically different from any previous production Jaguar, it followed the basic lines of the racing D-Types. The exceptionally long bonnet housed flared-in headlights with glass covers and chromed curved surrounds with side lights/indicator units set underneath, wrapping around the wing corners. The mouth in the nose was fitted with a horizontal chromed splitter bar with a center badge. Two chromed quarter-bumper bars curved around with the bonnet line under the lights, with twin overriders at the center, meeting the mouth of the bonnet. For most markets, the registration plate was by way of a sticker attached to the leading edge of the bonnet; where necessary a separate mounting plate under a bumper was used for some overseas markets. A bulge ran the center length of the bonnet to accommodate the camshaft covers of the XK engine. This was flanked by bonnet louvres either side to aid engine-bay cooling. A further vent exited from the rear of the bonnet's power bulge.

From the sideview, the extensive bonnet and integral wings appeared to form 50 percent of the entire length of the car. For the fixed head, doors still incorporated Jaguar's traditional chromed surrounds to the glass, and there was a large rear side window area formed to follow the line of the sloping

The famous E-Type three-spoke wood-rim steering wheel, only ever seen on the E-Type, formed a signature part of the car, even taking pride of place on the sales brochure.

roof and the upswing of the body over the rear wings.

At the rear, the fixed head sloping roof ran down to meet the rear light units, again curving around the sides of the wings above rear quarter bumps and overriders. The bumpers extended around the sides to meet the rear wheel arches. Registration plates were mounted centrally, set into the bodywork, and the large under-body valance extended up to meet it, under which the twin exhaust pipes were mounted together in the center.

Both fixed head and roadster were two-seaters, the only styling differences being that for the latter the door glass was frameless and there was a chromed strip at the top of the doors running

the width. The roadster also had a flat boot area with a rear hinged lid (the lock released from a pull lever inside the car behind the driver's seat). The boot interior was limited in space but was Hardura covered and housed underneath a spare wheel well, flanked by the fuel tank. A single-piece hood, attached to the rear, could be easily erected or taken down in one operation from the driver's seat. When raised, it secured to the windscreen on each side. Jaguar supplied a tonneau cover for the hood when down providing a cleaner line to the car.

Externally, nothing was carried over from the outgoing XK150. The E-Type was significantly lighter, the fixed head 24 hundredweight compared to 29, a

couple of inches shorter and 7 inches lower.

With everything new externally, the E-Type was a 50/50 mix of old and new under the skin. It retained the XK engine, now in 3.8-liter S form, complete with triple SU carburetors. A new drum air cleaner arrangement was fitted. Cooling was provided by the front-mounted radiator, but due to height restrictions, there was a separate header tank and the engine-mounted cooling fan was replaced by a thermostatically controlled electric fan. The E-Type also carried over the four-speed Moss gearbox from the XK. What was new, however, was an incredibly sophisticated independent rear suspension.

The whole of the rear suspension, plus the braking system and differential, were fitted to a detachable steel cage frame attached to the body by four rubber mountings. The design incorporated two quite small coil spring/damper units on each side, mounted on projecting lugs on lower suspension links, located at the top on the subframe. Large diameter half-shafts extended from the differential to the wheel hubs in light alloy castings, forming the top link of the system. These castings continued down to pivots on which were attached the lower suspension links, with forked ends, one attached to the wheel hub casting and the other to mounts at the bottom of the cage. A Salisbury rear axle was still used and solidly fixed to the subframe. Only two suspension members were not directly part of the cage frame—"U" section trailing radius arms, which located the suspension longitudinally and were fixed to the body with rubber mountings.

For the rear braking system, Jaguar fitted 10-inch inboard discs mounted directly on the output shafts of the differential, with special silicon rubber oil seals to guard against the extreme heat that could be generated in that area. There

The E-Type's extremely complex and well-executed monocoque was a major move forward for Jaguar.

was a single caliper per disc, operated by two pistons. The handbrake operated on an extra caliper on one rear disc.

The whole independent suspension subframe was a masterpiece of engineering and could be easily detached from the car for major work. The design first seen in the E-Type went on to appear in future Jaguar models for many years.

Returning to the braking system, conventional outboard discs of 11-inch diameter were fitted at the front. The brake pedal was connected by a balance bar to two master cylinders, each with its own fluid reservoir, and also to the brake booster bellows unit (servo). This was an unusual system designed by the Kelsey-Hayes company in America and adapted and made in the UK by Dunlop; its first car installation was in the E-Type. Depressing the brake pedal activated the master cylinders, simultaneously closing an air valve and opening a vacuum valve in the servo, vacuum being taken from the engine's induction system. The unusual bellows then started to contract, resulting in a load

applying through the balance bar to the master cylinders.

The E-Type was only available with 72-spoke wire wheels of a smaller 15-inch diameter, fitted with Dunlop tires.

Internally, the E-Type was again a mix of old and new. The dashboard and instrument layout first seen in the Mark 2 saloon was continued, and like the XK150 before it, there was no wood to be seen anywhere. Instead the E-Type relied primarily on black finishes for the top rail and surround, and black crackle finished metal for the dashboard area, except for the center section containing the auxiliary gauges and switchgear, which was finished in a figured silver alloy. This center panel was flanked by two slider controls, one on the passenger side for the heater control, the other on the driver's side controlling the manually operated choke. On the passenger side, there was an open glove box and a grab handle built into the top rail.

A center console ran down vertically from the dashboard and horizontally between the front seats featuring the

gear lever, radio controls, and ashtray. The sides were trimmed to match the rest of the car and the top in figured silver alloy. There was a "fly-off" handbrake fitted to the passenger side (right-hand drive cars) of the console.

A new style of steering wheel, only ever used on the E-Type, became a styling feature, even on the front cover of the original brochure. Held by three spokes, it was a drilled design in alloy with a wood rim and center horn boss featuring a checkered flag. The steering column was adjustable for reach and contained the usual stalk to control the flashing indicators and headlight flashers.

New style bucket seats were fitted, and the car was trimmed to the usual Jaguar standard with quality carpeting. The fixed head coupe headlining was now foam backed.

The launch of the E-Type at Geneva was a hasty affair, but the panic was well worth the end result as the orders rolled in and Jaguar was in the embarrassing position of being unable to meet the incredible demand. Not only was the

The original 3.8-liter roadster with the hood erect and displaying whitewall tires and wire wheels, common on cars destined for the US. The early 3.8-liter cars had external bonnet locks on each side that required a T-key to unlock, shown here covered with chromed escutcheons.

An early 3.8-liter fixed head coupe, arguably the prettiest of the two styles with its swept back roofline.

One could say that the E-Type was one of the first "hatchbacks" with this very useful luggage area in the fixed head and the wide-opening rear door. The spare wheel, tools, and fuel tank are underneath the boot floor and are easily accessible from a lift-up panel.

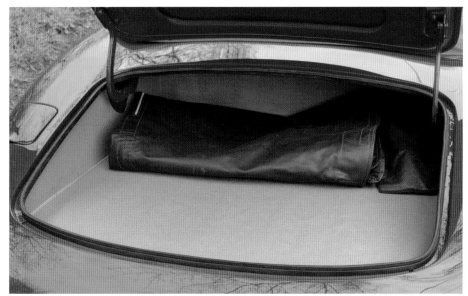

In contrast, luggage accommodation in the roadster was minimal and once you included space for the hood tonneau cover (seen here), you really needed an after-market luggage rack on the boot lid!

The legendary XK six-cylinder in triple carburetor E-Type form. The forward hinged bonnet provided excellent access for maintenance. At the bottom left, near the firewall, can be seen the bellows unit for the Kelsey-Hayes brake servo.

THE LIGHTWEIGHTS

After some initial success in competition, Jaguar decided to assemble a small run of E-Types specifically designed for racing. To reduce weight, the cars were produced in aluminum with only the front framing carrying the engine and suspension, still in steel.

Production E-Type roadsters could be fitted with a fiberglass hardtop at extra cost, and the factory determined that the racing lightweights would need a hardtop, not least to assist rigidity of the whole structure. These were also made of light alloy.

The front and rear suspension were modified, made stronger and lighter, and the braking system was also upgraded. Instead of wire wheels, Dunlop light alloy peg drive wheels were fitted, similar to those used on the D-Type sports/racing cars in the 1950s.

The engine was the most powerful version of the XK unit yet. Utilizing an aluminum cylinder block and fitted with fuel injection, the internals were also strengthened and fitted with dry-sump lubrication. With a racing "wide-angle" cylinder head, the new performance engine could develop just short of 350 brake horsepower. A German five-speed ZF gearbox was used instead of the Jaguar four-speeder. Just 12 Lightweight E-Types were produced.

E-Type so special in terms of its looks and performance, but at an original cost of just over £2,000 including UK taxes, there was nothing in the world that could touch it for value for money. When the road testers got hold of the car, they sang its praises as not just a highly credible and fast sports car that handled so well, but entirely practical for everyday use—quite happy sauntering along in dense traffic one minute, and the next accelerating from 0 to 60 miles per hour in under 7 seconds without hesitation.

1962–1963: Minor Improvements

In 1962, driver/passenger accommodation was improved. The footwells were modified to provide extra space, the rear bulkhead was altered to provide greater seat travel, and the foam seat fillings were thicker to give greater support, plus the position of the pedals was altered.

The only changes mechanically involved the fitting of shields over the rear universal joints for protection, increased thickness of the brake discs and new brake pad material.

From a starting sales figure of 2,182 in 1961, 6,266 E-Types came off the assembly line in 1962 and 4,065 in 1963.

1964–1965: More Flexibility and a Tidy-up

Some major mechanical upgrades took place towards the end of 1964 that affected not only the E-Type but the big brother Mark X as well. The most significant was a 4,235cc version of the XK engine. The increase in capacity was achieved with a new engine block, the two middle cylinders moved closer together and one and six moved further out and the bore increased by 5.07 millimeters. The block remained of the same dimensions.

The 3.8-liter "cockpit" area with instrumentation similar to the saloons but a distinct lack of the wood veneer for which Jaguar was well known. Black and alloy was considered sportier. This later 3.8 shows one of the chromed internal bonnet locks forward of the front door area. Although a very well-equipped car, a radio was still extra in those days!

Very early on, the E-Type was campaigned in racing. Even later, world champions like Jackie Stewart cut their teeth racing them, in his case, "borrowing" a new E-Type from his parents' showroom to compete!

(ABOVE) The 4.2-liter E-Type Series 1 fixed head was virtually indistinguishable externally from the early model.

(RIGHT) The interior of the 4.2-liter E-Type with larger, more comfortable seating, black finish replacing alloy, and even a center armrest/glove box on the console.

There was also a new crankshaft with thicker webs for the extra torque and a new vibration damper. The water jacket was altered to improve flow of coolant. New pistons with chromed-plated tops and oil-control rings improved oil consumption over the 3.8-liter engine. The induction manifold was cast as a single piece and the exhaust system was also revised. The main advantage of the new engine was to increase the torque figure by 43 pounds-feet, which improved mid-range flexibility.

There were other, more minor changes in the engine bay. An alternator replaced the dynamo for better charging and the radiator was made of copper instead of alloy.

With the 4.2-liter engine came a new four-speed all-synchromesh gearbox with inertia-lock baulk rings preventing the engagement of a gear until the synchromesh was complete. This was a significant improvement noted by owners as it allowed smoother changes, with less noise and removed the long "throw" from first to second gear of the old Moss gearbox.

The Kelsey-Hayes brake booster system was abandoned in favor of a conventional Lockheed brake servo, and the front discs received heat shields. With better sealing, the steering and front suspension didn't need so much maintenance, and mileage suggestions increased from 2,500 to a massive 12,000 miles.

Externally, the new E-Type was only identifiable by the "4.2" badging at the rear and for the keen-eyed, sealed beam headlights were now fitted.

Inside there were more changes. Firstly, the bucket seating had gone (often criticized for not being that comfortable), replaced with more conventional, flatter seats, also now adjustable for rake. The perhaps rather "garish" silver alloy finish on the center dash panel and transmission tunnel was replaced by black vinyl and leather.

MERGING AND AMALGAMATION

Jaguar had been an independent company, singularly controlled by Sir William Lyons since 1935. Through the 1960s, the company had expanded into a major group in the motor industry, owning companies like Daimler, Henry Meadows, and even Coventry Climax at one time. Times were changing in the automotive industry and the need to bring companies together for economies of scale was a major consideration.

Although several approaches had been made over the years, the final outcome was a merger between Jaguar Cars Ltd and the British Motor Corporation (BMC), who were responsible for a large proportion of the British motorcar industry: Austin, Morris, Riley, Wolseley, and MG. The new company was known as British Motor Holdings Ltd (BMH) and the merger took place in July 1966. Jaguar and Sir William Lyons retained their autonomy, but hopefully would gain the advantage of scale in the supply of components.

In less than two years, by May 1968, the situation had become more complex. Leyland, another big player in the British car industry with Triumph and Rover came together with BMH to form British Leyland (which included Jaguar), although the latter remained independent in most areas for a few years to come.

There was now also an armrest/glove box on the console between the seats. For the roadster model, the boot lid could be locked.

There was only a slight increase in the price of the 4.2-liter car over the equivalent 3.8-liter model. The new car was 1 hundredweight heavier and the performance was virtually the same. The 3.8-liter model did continue in production throughout 1964 until the introduction of the 4.2, the total number of the earlier models being 15,481.

During 1965, little changed except for the availability of radial ply tires, and water sealing in various areas was improved. There was also a new windscreen washer system.

1966: Getting a Stretch

March 1966 brought a third variant on the E-Type theme. Alongside the fixed head there was now a 2+2 model. This had the advantage of widening the customer appeal for the E-Type and allowed the fitment of an automatic transmission for the first time.

The change was achieved with surgery to the bodyshell, adding 9 inches to the wheelbase and 2 inches to the overall height. The extra length in the body had all been achieved in the mid area. The doors were now 8½ inches wider with a larger glass area, chromed surround, and even a horizontal chrome strip across the top finishing at a spear on the rear wing. To achieve the extra height in the roof, the windscreen was more severely raked. The overall effect was very well executed, and the car still didn't look out of proportion.

Changes internally amounted to the fitting of a rear bench-type seat upholstered to match the front. Legroom and width were limited so these were very much "occasional" seats but served a useful purpose for those with a growing family or when necessary to carry adults. The seat back could fold down

A new third model introduced in 1966, the 2+2. The "stretched" body and raised height have not compromised the style of the car. This is an early preproduction example without the chromed stripe above the door, below the window level that appeared on all production 2+2s.

This illustrates the differences between the Series 1 fixed head and 2+2 models.

(ABOVE) The 2+2 model was well received in the States as it was the first E-Type to be equipped with automatic transmission.

(LEFT) The occasional rear seating made the E-Type more appealing to families and those seats could be folded down to provide extra luggage space.

to allow for extra luggage in the rear. A benefit for the front seat occupants was that the seats could now travel further backwards when no one was sitting in the rear. The 2+2 model also featured a full-width parcel shelf below the dashboard and there was a larger glove box with a lockable lid. The heater system now had variable direction outlets and the doors had burst-proof catches.

The new version weighed in at an extra 2 hundredweight over the fixed-head. Minor mechanical changes were necessary, which included revised spring and damper rates for the extra weight. Internal cooling was improved with better heat shielding underneath the car.

For the first time, an automatic transmission was available on an E-Type due to the extra length afforded by a 2+2 bodyshell. The control occupied the center console just like a conventional gear lever, with both D1 and D2 operational modes. The 2+2 of course was still available with the manual gearbox as well.

1967: Halfway House

By the end of 1966, over 13,000 E-Types had been sold worldwide and a facelifted car was already under development at Jaguar. There would be a brief period during 1967 when some design features of the E-Type would change unannounced before the new model would be seen the following year. Effectively these were still 4.2-liter (Series 1) models in all three body types, but afterwards became unofficially known as Series 1½.

Because of the need to improve the rather dim lighting of the E-Type, this model now displayed open headlights (without the covers). This meant a new chromed surround and finisher. Other identifiable changes included door-mounted rearview mirrors and non-eared spinners retaining the wire wheels. This was a move brought about by overseas legislation.

Interior trim changes came about in dribs and drabs. Some of these interim cars featured new rocker switches on

the center dashboard panel replacing the old toggle switches, and with flush fitting choke and heater controls. The cigarette lighter was repositioned, and the indicator stalk activated the horn as well. Some cars featured different door panels and seats with adjustable rake and adaptable to take head restraints.

Some of these "Series 1½" models also featured engine modifications to suit the North American market.

To comply with new emissions regulations, considerable changes were made to the engine. The most economic and cost-effective way to modify the exiting 4.2-liter engine was by fitting twin Zenith-Stromberg 175CD2 carburetors. There was also a new inlet manifolding arrangement crossing over the top of the engine and connected to the exhaust manifold. The way the system worked was with a small-diameter choke tube. At low engine speeds, the mixture was ducted through the new manifolding, heated by the exhaust, and then fed back to the inlet side.

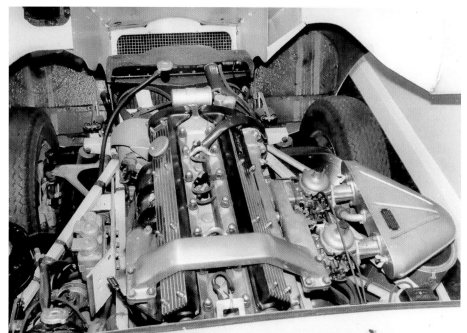

(ABOVE) An interim car that later became unofficially known as the Series 1½ model. Specification varied according to what was on the production line at the time, but all featured the open headlights with new surrounds, but retained the above bumper side/indicator lighting.

(RIGHT) E-Type engine equipped to meet US federal emissions regulations with the cross-over trunking and two Stromberg carburetors.

1968–1970: Enter the Series 2

Jaguar's E-Type Series 2 was announced at the October 1968 British Motor Show and there were lots of changes, some of which have been identified previously with the interim Series 1½ models.

Stylistically the Series 2 featured a new bonnet with a much larger (by 68 percent) mouth area to eliminate cooling problems in hotter climates. This necessitated a new chromed horizontal cross bar in the mouth. The open headlights mentioned earlier were a part of the new package, but they were fitted 2 inches further forward to minimize scatter, so the surrounds and fillet above the lights was different again to those seen on

the Series 1½. New lighting treatment continued with much larger side light/indicator lens units set below the bumper bar on a revised bonnet valance, with side repeaters for most markets fitted to the side of the front and rear wings. The bumper bars looked, but weren't, the same and the bumper heights were changed to suit new legislation.

At the rear, the bumper bar continued to a center section with the registration plate now fitted below on a satin alloy finished panel with reversing lights either side. The rear lighting units also formed part of this under-bumper panel of a new, larger design. To accommodate

The Series 2 model with larger air intake mouth, open headlights with different surrounds from those on the Series 1/2, and larger side/indicator lightings below bumper level.

a square type of registration plate, the exhaust pipes were more widely separated.

From the sideview, nothing had changed except for the 2+2, which got a new wraparound windscreen with an increased rake to the vertical; the bottom of the screen came further out taking up much of the space previously occupied for the scuttle area, and there were now only two wipers fitted to this model,

(ABOVE, LEFT) Series 2 cockpit area with a nod to safety with extra padding, recessed door handles, and rocker switchgear.

(BOTTOM, LEFT) Rear styling of the Series 2 with full-width bumper bar, under-bumper lighting attached to an alloy finished panel, and revised exhaust pipe exit. This is a 2+2 model.

with a stronger motor. The screen change enhanced the look of the 2+2. A driver's door rearview mirror became a standard feature. Although still available with wire wheels, for the first time E-Types were offered with 15-inch steel wheels as well, either in silver finish or chrome plated, fitted with standard Jaguar hubcaps but with black centers.

US federal regulations dictated changes to the interior. The rocker switches replaced toggles, there was a combined ignition and starter switch, and flush-fitted choke and heater con-

trols. The door trims were new, simpler in design in a recessed area to accommodate the new style flush-mounted door handles, and there were rounded knobs to the window winders and better anti-burst door locks.

New seating with a positive locking system so they wouldn't tip forward in an accident were fitted along with new seat belts.

The braking system now incorporated three pistons at the front, two at the rear to meet the new US requirements for safety. The engine followed the practice

first seen on the Series 1½, standardized for all US-destined cars, while the UK and other markets still received the triple carburetor arrangement. All engines by now got the ribbed cam covers.

For 1970, the US emissions equipment was amended, whereby the crossover manifolding was replaced by a water-heated chamber. On part throttle, the mixture was warmed in the chamber and then fed to the inlet manifold. At higher speed, a bypass allowed the full flow of the mixture directly into the manifold. Another addition for the US

(**ABOVE**) A few E-Types, primarily for the US market, were fitted with an air conditioning system with the controls and outlets below the center dash area.

(**LEFT**) Series 2 engine bay with revised emissions equipment and ribbed cam covers.

market was a charcoal canister to absorb gasoline vapor.

To simplify manufacture later in the year, the twin Stromberg arrangement and manifolding was standardized for all markets, so no further triple SU carbureted cars would be made.

Production of Series 2 models ran out towards the end of 1970 with development finished on its replacement model, the Series 3. A total of 18,848 cars were produced with the roadster being the most popular, followed by the 2+2.

1971: Series 3—The Final Fling

By this time, the Jaguar E-Type was losing out, particularly in the States where homegrown V-8s were the order of the day and could deliver huge performance. Jaguar could no longer address this with changes to the XK six-cylinder engine.

Jaguar had previous experience with a V-12 engine, fitted to the XJ13 sports/prototype, and although it would have been much simpler and less costly to produce a V-8, Sir William Lyons was concerned about the refinement and wanted to create a major impact on the market. A road-going V-12 would give Jaguar a step-up from the numerous American V-8s and even out-shine European machinery from the likes of Ferrari and others.

The V-12 engine had a single overhead camshaft on each bank of six cylinders operating in-line valves working in a flat cylinder head, with the combustion chamber formed in the recessed piston crowns. The new engine had a capacity of 5,343cc with a bore

(ABOVE) The Series 3 roadster with its longer (2+2) wheelbase, flared wheel arches and revised frontal treatment with a larger air intake and, for the first time on an E-Type, a grille.

(RIGHT) Only the second new Jaguar engine since the end of World War II, the 5.3-liter V-12 was fitted to the E-Type Series 3 from 1971, requiring significant changes to the front subframe.

and stroke of 90×70 millimeters. The crankcase was carried well below the level of the seven-bearing crankshaft. Wet liners were used in the block with upper flanges machined so that they left a narrow water passage between adjacent liners in the hottest part of the cylinders.

With the advantage of an aluminum block and cylinder heads, this large engine was only 80 pounds heavier than the XK unit. For lubrication, there was a crescent-type oil pump incorporating an oil cooler. Chain-driven camshafts remained like the XK engine.

Carburetion again used the Zenith Stromberg 175CD SE carburetors, but now four of them, located in pairs outside the engine's "V" formation. The beautifully engineered throttle linkage picked up on all four carburetors with rods leading to a centrally mounted capstan and then on to the acceleration cable.

Jaguar was the first company to fit the Lucas Opus (Oscillating Pick up System) ignition, with an electro-magnetic pickup and solid-state electronics.

With a quoted maximum power of 272 brake horsepower at 5,850 rpm, and maximum torque of 349 pounds-feet at 3,600 rpm, the Jaguar V-12 was a fantastic marketing tool for the company. It was well engineered, exceptionally refined, and could be updated over time to a larger capacity or even to accept fuel injection.

The V-12 engine was longer and wider and slightly heavier than the XK, so the front subframes of the E-Type were redesigned, incorporating triangular strengthening plates at the top. The bulkhead onto which the frames were located was also strengthened and enlarged to take the extra weight and stress. Also, to handle the increase in torque there was a substantial tie-bar arrangement under the engine.

The front suspension was fitted with the anti-dive geometry from the

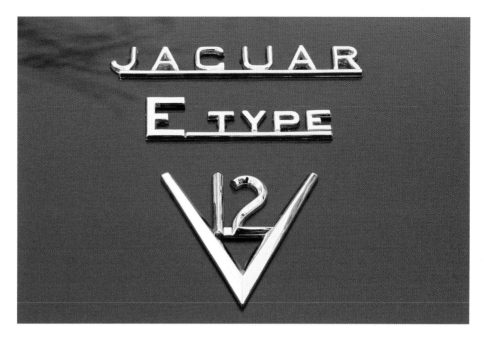

Jaguar proudly fitted this badging to the boot lid/tailgate of all Series 3 models.

XJ6 saloon by repositioning the upper and lower wishbone mountings, with the upper wishbones provided with sealed-for-life bearings. Torsion bar adjustment was now by means of a cam plate. The track was widened at the front by 4¼ inches and 3 inches at the rear. Wider 6-inch wheel rims were fitted with Dunlop tires.

The braking system was virtually the same as the Series 2 models but with ventilated discs at the front. The rear brakes were still mounted inboard, but cooling was improved with the fitment of an under-floor air scoop.

Power-assisted steering was now standard, and the rack was modified to reduce compliance. Jaguar four-speed manual transmission was still available with a larger 10½-inch-diameter clutch. Automatic transmission was by BorgWarner but now the Model 12 box instead of the Model 8.

Bodily, the new V-12 models used only the 2+2 shell and so were effectively long wheelbase cars. There were therefore only two models, the fixed head 2+2 and the roadster, still just a two-seater. At the front, the mouth was enlarged yet again, now incorporating a grille with a

chromed surround. There was also an air scoop underneath to move more air into the engine bay for cooling.

From the side, the wheel arches were flared to accommodate the wider track and tires. Both cars used the wider doors with chromed strips at the top and the windscreen rake followed that of the Series 2.

At the rear, the fixed head's tailgate displayed a prominent "V-12" badge and chromed vent panel. The rest of the rear below bumper level was as the Series 2, although the V-12 now sported a four branch "finned" tailpipe arrangement.

The interior was still very much existing E-Type. The floor was lowered slightly to improve legroom, and the seat backrests reclined further and got headrests as standard. For the first time since the E-Type was launched in 1961, there was a new steering wheel of a smaller diameter and leather trimmed instead of wood. It retained the three-spoke arrangement but was now dished for safety, with a simpler embossed growler featured on the black center boss.

Having the wider opening doors made entry and exit easier, particularly for the roadster with the top up.

(ABOVE) A Series 3 with fitted hardtop and conventional steel wheels with hubcaps.

(RIGHT) For the US market, E-Types had to be adapted with Nordel bumper over-riders.

(LEFT) A Series 3 interior, with a parcel shelf below the dashboard and the new style steering wheel.

THE SERIES 3 SIX

Despite the changes and launch of the V-12, Jaguar had intended to still produce a Series 3 with the six-cylinder XK engine. The original sales brochure confirmed this as both engine models were displayed and featured in the text and specifications. However, only a mere handful of the Series 3 sixes were actually produced, some of which still survive to this day. Jaguar had decided that the XK engine could not produce the required performance in the new car.

Jaguar also produced a revised fiberglass hardtop as an extra-cost option, with a larger rear screen and chromed side ventilator panels.

Although a few Series 3s came off the production line in 1970, full-scale production started early in 1971 with 3,833 available for sale, nearly twice the quantity of either the Series 1 or 2 in their first full year of production.

1972: More of the Same

For 1972, several minor changes took place to aid production and keep costs down. The handbrake assembly was changed so that the same parts could be used for left- or right-hand drive models. A revised starter motor was fitted and a simpler coolant hose system and water pump. Later in the year, a sealed fuel system was added.

On the comfort side, the heater and choke controls were redesigned and fresh-air vents with control levers were fitted. To meet new safety legislation in the States, a seat belt alarm system was fitted and rear seating in the 2+2 had to be fixed in the upright position. There was now remote-control operation for the door rearview mirrors.

It was good year for Series 3 sales with another 4,000 leaving the production lines.

1973: Fixed Head Finale

The big news for 1973 was the deletion of the fixed head model due to safety legislation. New US law required coupes be equipped with rollover bars, which could not be factory fitted by Jaguar. Surprisingly, although the model had often been criticized for its styling with the wider wheel arches, the fixed head outsold the roadster with a total of 5,789 produced, compared to 2,058 roadsters up to that point.

Roadster sales continued. Jaguar had already been offering an air conditioning system for US customers and this was made available for the home market.

Minor changes during the year included revisions to the rack-and-pinion steering assembly, the fitting of a thermostatic vacuum switch to the right-hand rear coolant branch pipe eliminating associated hoses, and new needle roller bearings in the gearbox.

Addressing US federal emissions regulations meant a loss of power for the V-12 engine. A new air injection system

The Series 3 fixed head was a short-lived model.

using a rotary-vane pump driven by a V belt off the water-pump pulley pushed air through the individual air injection pipes into the exhaust gases to ensure efficient burning. There was also a new fuel filler dispensing with the fuel stop tap.

The cooling ducts for the rear brakes had to be modified, as they got too easily damaged; there was a modified BorgWarner automatic transmission, first seen in the XJ12 saloon; and there was a modified synchromesh for the manual gearbox.

1974: End of the Line

1974 was the final year of E-Type production. Despite this, January was a banner sales month with 580 cars, a record for Jaguar in a single month for any E-Type variant.

One of the reasons for the impending demise of the E-Type was more legislation coming from the States relating to safety and emissions. One such regulation meant that all cars had to sustain a 5-mile-per-hour impact at the front or rear. The only economic option for Jaguar was to fit US cars with Nordel rubber "buffers," extending forward from the framework, while at the rear adding a pressed steel reinforcing plate to the monocoque to distribute impact.

It was also during this final year of production that the exhaust system was amended with a twin "fin" tailpipe.

With the last E-Type leaving the line a total of 7,982 roadsters had been completed. It truly was the end of a very special era in Jaguar's history.

SPECIAL EDITION

The very last E-Type Series 3 models produced were special editions to commemorate the 14-year history of the model. Just 50 were made, 49 of which were finished in black with the other in British Racing Green. All roadsters, they had the chromed steel wheels with hubcaps and as a mark of their special place in the history of the E-Type, each dashboard carried a brass plaque inscribed with the car's chassis number and Sir William Lyons signature. They were sold at a price of £3,812 or about $9,837.

SPECIFICATIONS

MODEL	3.8 Series 1	4.2 Series 1	Series 1 2+2	4.2 Series 2	5.3 Series 3
ENGINE SIZE	3,781cc	4,235cc	4,235cc	4,235cc	5,343cc
CARBURETION	3 x SU	3 x SU	3 x SU	3 x SU (later 2 Strombergs)	4 x Stromberg
MAXIMUM BHP	265@5,500	265@5,400	265@5,400	265@5,400	272@5,850
MAXIMUM TORQUE	260@4,000	283@4,000	283@4,000	283@4,000	304@3,600
GEARBOX	4-speed	4-speed	4-speed	4-speed	4-speed
AUTOMATIC	n/a	n/a	BW 3-speed	BW 3-speed	BW 3-speed
0 TO 60 MPH	7.1 sec.	7 sec.	8.9 sec.	8 sec.	6.4 sec.
STANDING ¼ MILE	15 sec.	14.9 sec.	16.4 sec.	15 sec.	14.2 sec.
TOP SPEED	149 mph	150 mph	136 mph	150 mph	146 mph
AVERAGE FUEL CONSUMPTION	19.7 mpg	18.5 mpg	18.3 mpg	18 mpg	14.5 mpg

CHAPTER SEVEN

ONE-MODEL POLICY

The XJ Series

Jaguar's postwar success was attributed to a fine range of quality saloons and sports cars, developed through a racing heritage, all based around the XK twin-cam engine. The large saloons were the backbone of the range and justifiably carried the slogan "Grace . . . Space . . . and Pace." Although the Mark X from the 1960s had been relatively well received, it didn't generate the sales intended, so Sir William Lyons and his team set about creating what would be, for him, the pinnacle of his styling ability, the XJ.

To confirm this model's success, one only has to look at the number of years it was in production and the quantity produced, as well as the increasing interest in the model still today.

A Family Resemblance

The XJ6, at first glance, carried over many of the styling features of the Mark X/420G, but in reality, every single body panel was new. Lyons had created a more balanced look with larger wheels in proportion to the body.

A new anti-dive feature improved the ride. The upper wishbone pivot was inclined at 3.5 degrees to the horizontal and the lower one angled downward by 4 degrees, causing the king pin to rotate under braking and push the body up. The effect kept the car level on deceleration. Girling dampers were fitted outside of the springs.

The new front subframe carried the engine mountings and rack-and-pinion steering with Varamatic power assist.

Front brakes incorporated three-wheel cylinders operating two smaller pads on the outside of each disc with one larger pad on the inside. This improved pad life and reduced brake fade.

The conventional 4.2-liter XK engine powered the car, either with twin 2-inch SU HD8 carburetors or twin Stromberg 175CDs in the US. The engine used a

Sir William Lyons was justifiably proud of his XJ creation, here seen at Browns Lane where the cars were produced at the time of the model's launch.

---

larger impellor in the water pump, and a larger bypass hose fitted in the cylinder head, eliminating the water gallery. A cross-flow radiator with separate header tank fitted behind the core kept the height down. The XJ6 bonnet had a bulge to clear the cam covers.

A new engine of 2,791cc (2.8 liters) was introduced for the XJ6, with a bore and stroke of 82×86 millimeters. This configuration would help sales in some European countries where there was a tax tied to capacity. With a cylinder head

RATIONALIZATION

The major motivation behind the design and introduction of the XJ6 was to enable Jaguar to rationalize its saloon car range. From nine cars in 1966 (Mark 2 2.4, 3.4, 3.8, Daimler 2.5-liter V-8, S-type 3.4, 3.8, 420, Sovereign, Mark X), the saloon range was winnowed down until the "new" XJ6 was the only model surviving by 1970, apart from the Daimler DS420 limousine.

The new XJ6 had therefore become a "global" car for Jaguar, satisfying all requirements for a luxury saloon.

identical to the 4.2-liter engine, it too featured the twin SU carburetors.

For the manual transmission models, Laycock de Normanville overdrive was again offered and the 4.2-liter engine retained the BorgWarner Model 8 auto-

matic transmission; the 2.8-liter engine used the smaller BorgWarner Model 35.

Wheels were 15 inches in diameter with 6-inch wide rims. Dunlop developed special E70 VR15 tires for the XJ, derived from their SP range.

(ABOVE) Instantly recognizable as a Jaguar, even though the radiator grille was somewhat nondescript, the lines and proportions were more popular than the previous model.

(RIGHT) The XK six-cylinder engine installation was very similar to that in the earlier 420, although the under-bonnet space had always been planned for a larger unit.

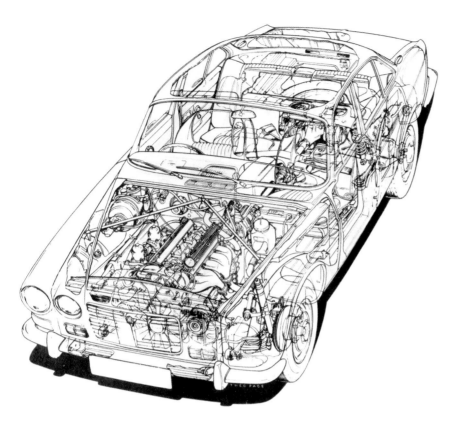

(LEFT) Main mechanical aspects were unchanged from the previous model—4.2-liter XK straight-six engine and independent rear suspension cage but with tweaks to the suspension and braking.

(BELOW) Larger wheels filled wheel arches better, while early cars bore no mention of the name Jaguar. The straight exhaust tailpipes were later curved to eliminate fumes entering the cabin when windows were open.

Although having that strong family resemblance to the Mark X, the styling of the XJ was less flamboyant, smoother, slightly smaller, and much lighter. With the roof supplying more strength, similar assembly methods were employed but with greater "crushability" in the design to better absorb impact in an accident. The front wings and lower rear wing panels were bolted for ease of removal and repair.

At the front was a similar four- head-light treatment with wraparound side lighting with mock horn grilles similar to the 420 and revised full-width bumper bar and overriders. The radiator grille was "egg-crate" style without a prominent surround, merely a jaguar-head design within the center badge. Gone was any recognition of the Jaguar name and, because of legislation, no mascot appeared on the bonnet. There was no center chrome strip along the bonnet. A large curved one-piece windscreen was raked quite sharply into the roof and from the side the whole window area was larger.

The sides were flatter with a waist swage line running the whole length of the body. Behind the rear door there was a discreet "kick up" in the panel. Window frames were lighter and of a simpler design and the door handles came from a standard supplier, common to many other British cars.

The boot lid was much flatter than before with a raised floor accommodating the spare wheel. The rear was more upright with entirely new lighting and a large under-bumper valance with cut-outs for the twin exhaust pipes.

Badging was restricted to a plastic leaper badge at the base of each front wing, the jaguar head in the grille and the engine size on the boot lid.

Internally, although the dashboard layout was the same as early 1960s Jaguars, the instruments were of a new,

The interior retained a traditional look with walnut veneer, yet better, more supportive seating, "safety" switchgear, and extra storage space in the doors. This early car with manual transmission features the chromed bezels to the instruments.

clearer style. Auxiliary gauges in the center were fitted on a black ribbed plastic backing set into the veneer. Rocker switches, first used on the E-Type Series 2, operated all the central controls, including the main lights. A set of warning lights was set within a binnacle between the speedometer and rev counter. The dashboard had a padded top roll in black.

Below dash level was a parcel shelf on each side with occasional tray in the center. The center console contained heater/air conditioning controls and radio, along with gear lever/transmission selector, two chromed ashtrays and cigar lighter, all set into a figured alloy backing. A lockable glove box on the passenger side of the dashboard and a center-console storage area within the armrest provide secure storage. Finally, there was a new steering wheel with twin spokes and half horn ring.

Seating was new and smaller, providing better lateral support and Posture Springing. The front seats reclined. A bench seat served rear passengers. Foam-backed headlining was used, and new door trims incorporated a slim wood fillet and combined armrests/storage areas. Recessed chromed door handles were used, plus either hand window winders or switches on the center console and at the rear for individual occupants if the car had power windows.

Some differences were tied to the engine chosen. The standard-trim 2.8 used Ambla instead of leather upholstery, with no rear-seat armrest and no rear-door storage pockets. Power assisted steering was not fitted. But one could specify a De Luxe version of the 2.8, appointed like the costlier 4.2-liter model.

The most prominent change to the interior came with the entirely new heating/ventilation and air conditioning system. Designed by Delaney Galley, it provided substantially greater output and airflow than any previous system used by Jaguar. Better temperature regulation was provided via controls on the center console and stale air was extracted via a one-way vent in the rear parcel shelf. A combined air conditioning option was available at extra cost that didn't require any substantial changes to the other system or take up extra space. All cars had fresh-air ventilation outlets, one each side of the dashboard.

1968: Launch Time

The XJ6 was launched in September 1968 with its first public appearance at the British Motor Show a month later. Priced at a mere £1,797 including tax for a 2.8-liter model, and £2,397 for a 4.2-liter automatic, these represented exceptional value for money. Comparative examples included the Mercedes-Benz equivalent at £3,000 more and the Oldsmobile Cutlass at £1,200 more.

Despite the marketing plan to sell the 2.8-liter as a base model to attract business from other lesser brands and to help tax conscious markets, the car never sold that well. Jaguar never even offered an example for the press to road test.

1969: Daimlerization

Gas fumes were an initial common problem, apparently caused by twin fuel fillers. An antisurge flap in the fillers solved the problem. The cars were also improved with heat shields fitted between the exhaust and the body under the front seat areas to prevent heat build-up. Quieter running camshafts were fitted to all engines from the end of the year. Stiffer springs and revised anti-roll bars and mountings provided a better ride.

Inside, front seat backs were modified to take optional headrests, and inertia reel seat belts were offered also as another extra-cost option.

The biggest news in 1969 was a new model in the XJ range, the Daimler Sovereign. With the run down in production of the Daimler 420 Sovereign in August, its replacement was announced in October.

The new model was identical to the Jaguar XJ6 externally except for a traditional fluted radiator grille, similar treatment on the boot lid and badging. Internally, it was also much the same except for the black fabric finishing to the center console and more badging.

At a marginally higher price, the Daimler was supplied standard with certain options one paid extra for on a Jaguar, like overdrive.

Good sales totaled 13,172 Jaguar XJ6s and 597 Daimler Sovereigns for 1969.

2.8 ENGINE FAILURE

It wasn't long before owners of 2.8-liter-engined cars complained of engine failure due to holed pistons. This only came to light when the cars were in road use and was found to be caused by a build-up of carbon debris on top of the pistons with the exhaust valves being in too close proximity.

This matter never occurred during Jaguar testing of the new engine because of the greater engine speeds achieved preventing the build-up. In customer use, the cars were driven more sedately, allowing the debris to build.

Jaguar later offered revised pistons and suggested the resetting of the static ignition timing. Subsequently the engine was dropped from production.

Series 1 models in both Jaguar and Daimler forms, the latter with the more prominent radiator grille and center chromed strip along the bonnet.

1970–1971: Ongoing Improvements

The XJs proved to be a phenomenal success and sales in 1970 for all models amounted to a staggering 21,833. By the end of 1971, another 27,517 XJs had been sold, resulting in an amazing 63,758 in a little over two years, a record for the company.

The cars were so good that few changes were necessary. Despite modified spring pans being fitted to the front suspension early on in an attempt to provide better tire clearance, it was only when the front wing wheel arches were modified in 1970 that the problem was cured.

Front seat occupants now had controlled fresh-air vents, supplied by air taken in from the outer headlight surrounds. The surrounds to the instruments, previously chromed, were from 1970 finished in matte black to avoid glare, accompanied by a satin finish to the exterior scuttle ventilator (replacing chrome). Aluminum door treadplates were also fitted and there were improved door locks.

Mechanically, emissions exhaust manifolds were now fitted to all engines along with an improved crankshaft rear oil seal. The brake fluid reservoir was also repositioned.

The biggest single change in 1970 was a new BorgWarner Model 12 automatic transmission offering a D2 position and part-throttle down changes without flooring the accelerator.

External changes meant new lighting to suit European regulations and a three-piece rear bumper to replace the previous one piece to save on replacement costs.

1972: 12 into 6 Will Go

The XJ body had been designed to accept another Jaguar engine, the V-12, first seen in the E-Type Series 3. As that engine was not yet in production during the XJ's development, many adjustments were necessary to fit the much bulkier engine and its ancillaries in the bodyshell.

Part of the problem was excessive heat generated in the engine compartment to the point that the battery, situated at the firewall, had to be fitted with its own cooling fan! The crossflow radiator was redesigned with two areas to ensure an even temperature across the V's two cylinder banks. An electric fan supplemented the conventional engine driven fan. Heat shielding was provided to protect areas like the steering rack and engine mountings, and the exhaust downpipes were double skinned to reduce noise. Heat shields were even fitted to the underbody to reflect heat generated by the exhaust system.

The V-12s were available only with the new BorgWarner Model 12 automatic transmission.

Front spring rates were increased due to the extra weight of the engine and its ancillaries (129 pounds). For a more efficient braking system, ventilated front discs were fitted with a Girling Supervac servo with an additional vacuum reservoir. Dunlop's special tires were upgraded for all V-12 cars, incorporating a steel breaker strip because of the car's weight and speed potential, fitted to ventilated road wheels.

Externally, apart from the revised wheels, changes were minimal. The "egg-crate" radiator grille was replaced with a more attractive one with vertical slats and incorporating a "V12" badge. On the boot lid, another "V12" emblem appeared and for the first time a Jaguar badge was fitted on the opposite side of the number plate area.

Inside, subtle changes appeared. Like Daimler six-cylinder models, a black finish replaced alloy on the console, which now incorporated a gold finished "V-12" badge. Similar door trims with pull handles were also used. Calibration of the main instruments changed, including a new 7,000 rpm rev counter. There was also a manual choke control fitted under the dash near the driver's door.

The XJ12, as it was called, was launched in July 1972. Shortly afterwards, the inevitable Daimler-badged version, the Daimler Double Six, was announced. In September, yet another model was announced, the Daimler Double Six Vanden Plas.

The "VDP" was a stretched-bodyshell version of the XJ, lengthened 4 inches for more rear-seat room and with wider rear doors. The model was built as a conventional Daimler Double Six at the Jaguar factory, without final paint or interior trim. The semifinished cars were then shipped down to the Vanden Plas Coachworks in London, where they were treated to final painting in a bespoke range of colors, fitted with chrome trim including auxiliary lighting at the front and a swage line, given "Vanden Plas" badging on the boot lid, and received a vinyl roof covering.

Internally, there were bigger improvements. The VDP model was fitted out more luxuriously with the best leather trim throughout to a different design, individual rear seating, and burr walnut veneer with boxwood inlays.

The most expensive model in the Jaguar/Daimler range, the Vanden Plas cost £5,363 at launch.

Later that year at the British Motor Show in October, the longer wheelbase bodyshell from the VDP became an option on other XJ models.

(LEFT) Apart from badging and the different profile tires, the XJ12 was identified by its vertical slatted radiator grille.

(BOTTOM) Even though the under-bonnet area was "made" for the V-12 engine, it was a tight installation. Note on the top right the cooling fan in the battery casing to dissipate heat and the well-engineered throttle capstan unit in the center of the "V."

The launch of the XJ12 at the London Motor Show, then the fastest production saloon car in the world.

1973–1974: More Than a Facelift

In September 1973, the Series 1 cars, as they later became known, were replaced by the Series 2 XJ. The public had purchased over 98,500 of the first iteration, an absolute record for Jaguar at the time. Jaguar announced the new version at the 1973 Frankfurt Motor Show.

Stylistically they were instantly identifiable from the raised bumper bar level at the front, necessary to meet new US regulations. This meant new "underriders" instead of overriders with a larger air intake below bumper level in the valance. Accompanied by new under-bumper side/indicator light units, the radiator grille was much shallower—and fluted for Daimlers.

The front wings and bonnet were adapted from the Series 1, but the doors and rear wings were unchanged. Rear lighting was revised.

The major change not easily noticeable was a substantial reworking of the firewall area, mostly to accommodate a new heating/air conditioning system. The normal double-skinned metal bulkhead was replaced by a single skin covered on the engine side by a full-width asbestos shield, and on the cockpit side with a mix of bitumen, felt, and PVC shaped coverings. A light alloy pedal box was fitted, sealed to reduce the ingress of heat and noise. Also, all drilled holes through the bulkhead were eliminated, replaced by multipin connectors taking the wiring connection from the engine bay through to the interior. Even the heating and air conditioning liquids were accommo-

THE VANDEN PLAS CONNECTION

(TOP) The top-of-the-range Daimler Double Six Vanden Plas, produced in low volume and finished off by the coachworks in London. The look was unique at the time with vinyl top, swage line chrome trim, bespoke paintwork, and even fitted auxiliary lighting.

(BOTTOM) The interior of the Vanden Plas with burl walnut veneer, boxwood inlays, and different seat design. This car is equipped with the extra-cost option of dictation equipment.

The Vanden Plas coachbuilding company was founded in Belgium. Their designs for early motor cars were well received, especially in the UK. Initially working as an agent in London, the company was reformed by the early 1900s into Vanden Plas Ltd., importing VDP bodies from the continent. Soon after production commenced in the UK, the company was preparing bodies for prestigious manufacturers like Bentley and Rolls-Royce.

After World War II, the business moved into the hands of the Austin Motor Company, later to be the British Motor Corporation. Later still, BMC was integrated with Jaguar to form British Motor Holdings and then British Leyland, incorporating the majority of British car manufacturers.

Vanden Plas produced several up-market versions of contemporary saloons, including the Daimler-badged Jaguars. The Daimler DS420 limousine was built and trimmed at the Vanden Plas coachworks in London until the business was closed down, after which the DS420 operation moved to Jaguar's Browns Lane factory in Coventry.

With the demise of British Leyland, Jaguar retained the rights to use the Vanden Plas name for up-market versions of their saloons, and also to identify some export models, like those for the States, that would normally have been named Daimlers.

(ABOVE) The XJ Series 2 model easily identified from the front by the raised bumper bar treatment and new narrow grille. This is a V-12 model with appropriate badging and extra-cost auxiliary lighting. The leaping mascot was not a factory fit.

(RIGHT) Rear view of the Series 2 with revised plinth now incorporating the registration lighting (previously in the bumper bar). New badging at this time incorporated "L" for long wheelbase where appropriate.

dated by sealed metal tubing set into the bulkhead.

The purpose of the changes was to insulate the new heating/air conditioning system from interference by external airflow and temperature conditions. The controls for the new unit were electro-servo operated.

Substantial changes took place internally, including a completely new dashboard layout with all instruments in front of the driver. A large air vent appeared where the minor instruments were before. Many of the old rocker switches were replaced with either push-button controls on the center console beside a new clock or by stalks off the steering column. The under-dash parcel shelf was replaced with a larger glove box and door pockets. The center console incorporated larger switches for the electric windows, where fitted. Another new steering wheel design appeared, now without a half horn ring. Door trims were also altered, the rear incorporating ashtrays. Fiberoptic lighting, a first for any British car, eliminated the need for some individual dash bulbs.

The new model range was extensive, composed of short- and long-wheelbase bodyshells for XJ6 and Daimler variants, long wheelbase only for the XJ12 and DD6 models, plus new two-door coupe bodies also in Jaguar and Daimler 4.2- and 5.3-liter models. Known as the XJ6C and XJ12C, these additional models were well received, but not instantly available.

Using the standard wheelbase shell, the new coupes had doors that were 4 inches longer with frameless windows and no central "B" pillar. With the front windows and rear side windows wound down, this created an attractive pillarless style. The coupes had a black vinyl roof covering hiding a change at the rear of the roof with extra strengthening, and other extra strengthening was fitted behind the door shut face.

A completely new dash layout for the Series 2 models with all instruments in front of the driver, a new steering wheel, and extra ventilation in the center panel.

A long-wheelbase Series 2 showing the extra legroom in the rear compartment. This car is equipped with the then new cloth upholstery.

(ABOVE) To many, the most attractive of all the XJ Series models, the two-door Coupe with wider doors, pillarless glass areas, and vinyl roof.

(RIGHT) Series 2 models destined for the States had to be fitted with federal-style bumpers incorporating 5-mph impact beams. Note also the name Executive, which applies to certain markets only and, in this case, the discreet British Leyland badge, not always fitted, except in cases where cars were assembled abroad.

1975–1976: Improved Economy, Performance, and New Models

As the 2.8-liter engine continued to give problems, Jaguar finally deleted this option and in April 1975 introduced a new 3.4-liter (3,442cc) version of the XK unit.

The "new" engine used the existing 4.2-liter cylinder block and the straight-port cylinder head. Fitted with twin SU HS8 carburetors with the usual automatic cold-start enrichment device, the engine was rated at 161 bhp. The new model was available with the usual choice of manual/overdrive transmission plus the BorgWarner Model 65 automatic, with a 3.54 rear axle ratio.

The 3.4-liter car was an economy model available only with the four-door bodyshell and various trim changes, like narrow pleated cloth-trimmed seating in a minimal range of just five colors. Unavailable in the US, the new model was competitively priced in the UK at nearly £400 less than a 4.2.

A Bendix fuel-injection system adapted by Lucas was fitted to the V-12 engines, initially for the two-door coupes only. This improved economy, helped reduce emissions, and raised the brake horsepower figure to 285 at 250 rpm less. An electronic control unit monitored data from various sensors around the engine and its ancillaries, computing the engine's fuel requirements, and sending the necessary information back to the injector solenoids.

In the engine bay, the installation looked more complex but was very efficient and easily maintained. At the same time, handling was improved with a revised steering rack, anti-roll bar, steering arms and upper wishbones. Finally, a rear axle ratio change to 3.07 improved economy.

Externally, a "Fuel Injected" badge appeared on the boot lid and a few

Series 2 XK engine with emissions systems.

The V-12 HE in the Series 2, now fitted with fuel injection.

RACING EXPLOITS

It was unusual for a large saloon to participate in racing, yet Jaguar, with the aid of Broadspeed Engineering, set about attacking the European Touring Car Championship in 1976/77 with race-prepared examples of the XJ12 Coupes. They had high hopes of promoting sales and beating arch-rivals BMW with their CSi models.

Whilst ultimately unsuccessful, the XJs competed in no fewer than eight races, leading them all at one time. XJs had also qualified fastest on five occasions, but only three times did a car finish the race.

months later fuel injection was standardized on all V-12-engined cars for all markets. There were also numerous trim changes throughout the life of the Series 2 cars. V-12s for example received the vinyl roof treatment and chromed swage line trim previously only on the VDP cars, and a new style Kent alloy wheel became available from GKN.

Another new model arrived on the scene, extending the Vanden Plas model offering with the 4.2-liter engine.

1977–1978: Final Changes and Improved Sales

For 1977, Jaguar replaced the aged BorgWarner Model 12 automatic with General Motors' GM400 transmission in the V-12 cars. The new gearbox provided sportier gear-changes and acceleration with lower gearing on first and second that could be held for longer.

In May 1978, all 4.2-liter engines received Lucas-Bosch L-Jetronic fuel injection with a three-way catalytic converter, turning pollutants into harmless elements. To combat power loss, the fuel injection raised the compression ratio to 8:1 and, in conjunction with larger inlet valves, provided an extra 15 brake horsepower, plus better fuel economy. Although the fuel-injected 4.2 was originally destined for the US market, this engine was standardized for all production cars in that year. The 3.4-liter engine continued with carburetion.

In October 1978, Rover's five-speed manual gearbox became available for the 4.2-liter XJ. This was a relatively new gearbox designed for Rover's SD1 saloon, revised to reduce noise and with a new outer casing to make it fit. It replaced Jaguar's now aging four-speed/overdrive unit but didn't much improve economy or performance.

The last Coupe models were phased out in November 1977 due to continued quality problems and capacity needed to meet demand for saloons. The Series 2 saloon continued to sell well into 1979 when its replacement was announced.

1979: An Evolutionary, Rather than Revolutionary, Change

Total Series 2 sales had surpassed 100,000 when the Series 3 was announced in March.

Although carefully retaining the essence of the XJ design, the third version saw significant changes.

The first obvious shift was in the bumpers, now wraparound rubber-faced steel structures, which for the US market incorporated 5-mile-per-hour impact beams.

The Series 3 XJ with its larger glass area, flatter roofline, flush-mounted door handles, new wheel trims and bumpers.

Comparison between the Series 2 (left) and Series 3 clearly showing the many differences in body and trim.

FROM JAGUAR CARS TO BMH TO BL

Jaguar Cars was an independent company until 1966 when an agreed merger took place between Jaguar and the British Motor Corporation (including Austin, Morris, MG, Riley, and Wolseley). The idea was to give British manufacturers improved profitability and technology. The new business became known as British Motor Holdings (BMH).

By 1968, BMH was combined with Triumph, Rover, and several motor industry suppliers and overseas interests, to form British Leyland. Having a troubled time as such a large conglomerate, Jaguar was finally privatized again in 1984.

A chrome blade sat atop the bumper and turn signals were into the front.

The radiator grille had a flatter, more prominent surround, vertical slats with a center rib, and displayed a jaguar-head badge. Daimler models retained their ribbed style with a "D" motif. The twin headlights, now quartz halogen for all but 3.4s, also incorporated the side lighting.

With the front screen and pillars more vertical, a flatter roofline and the rear of the roof raised slightly, there was a sleeker look to the car with extra headroom inside. Marginally reducing the roof width and leaning the slightly larger glass area inward improved visibility. All models but the 3.4 now featured tinted glass, and the front door window

quarterlights were eliminated. For the first time, Jaguar offered an electrically controlled flush-fitting sunroof.

Front and rear wings were subtly redesigned, and along the swage line there was a single coachline featured on XJ6 and a double on XJ12 models. The 3.4 was plain. The Series 3s now had flush-fitting door handles.

At the rear, the full-width wraparound bumper incorporated the rubber molding and inset high-intensity fog lights (not for US cars). Larger rear light clusters now incorporated the reverse lights and there was a new wider, flatter boot plinth incorporating the lock and illumination. Badging was amended according to model, but all of the same plastic, chromed faced style.

Road wheels remained of the same size but with a new style stainless steel trim with exposed wheel nuts. Alloy wheels were still an option on all models.

Although the basic elements of the mechanics of the cars remained the same, there were many refinements, like cruise control operated from one of the steering wheel stalks. Only available for the larger (4.2/5.3 liter) engines, cruise control was standard on US-bound cars.

Front seating was improved with lumbar support adjustment and power seats for the US market, optional elsewhere. New map pockets, hidden inertia-reel seat belts, and slightly altered rear seating improved comfort all round. A new molded headlining incorporated flush-fitted sun visors and there was a larger rearview mirror. Carpeting was also of a plusher type, which improved sound deadening.

Instruments had improved legends and all switchgear now incorporated symbols instead of words. The warning light display now incorporated extra warnings for low coolant level and fog lights on. Another warning light illuminated if there was a bulb failure. A new steering wheel provided better view of

the instruments and the column stalks were swapped to meet internationally recognized positioning.

New driver aids included intermittent wipers and a headlight wash/wipe system standard on VDP models and those for the States. Power door mirrors were also standard on those models. As a final touch, reminiscent of classic Jaguars, there was a fitted briefcase containing tools in the boot.

The 4.2-liter engines retained fuel injection with the OPUS electronic ignition system and now developed 200 brake horsepower. For economy, fuel supply was cut off when the driver released the accelerator until revs dropped to 1,200. The 4.2-liter cars could still be fitted with the ex-Rover manual gearbox or BorgWarner Model 65 automatic.

Series 3 V-12s were fitted only with the GM 400 automatic transmission.

1980–1982: Economy Is King

At launch, service intervals had been set at 3,000, 6,000, and 12,000 miles, but were soon increased to 7,500 and 15,000 for the US market, and for all markets by May 1980.

In October 1980, Jaguar announced a revised 3.4-liter XJ6 to help compete in the important fleet market. Reduced in price by £500, the car came with a cheaper audio system, cloth upholstery, and straight-grained wood veneer.

Due to a changing climate over fuel economy, the V-12 engine was receiving bad publicity for its inefficiency. Jaguar worked with a Swiss engineer, Michael May, who had some radical ideas on combustion chamber design. A two-chamber combustion arrangement was devised with the inlet valve recessed in a collection zone, and the exhaust valve located higher up in a "bathtub" chamber into which the spark plug projected. The mixture was pushed by

A later 3.4-liter Series 3 XJ6 with cloth upholstery and the straight grain woodwork. This car also features the later center-console area with revised clock mounting. Better equipped models featured a combined clock and computer.

the piston from the inlet valve zone to the combustion chamber on its compression stroke. This meant low turbulence, concentrated the charge around the spark plug, and enabled rapid complete burning of the lean mixture. The new cylinder head had a much higher compression ratio of 12.5:1 and a revised Lucas-Bosch D-Jetronic fuel injection system was adopted. Although overall performance was little changed, fuel consumption was dramatically reduced.

With this new engine, the cars were sold with "HE" appropriate badging. The V-12s also received standard alloy wheels, chromed side-moldings, electrically operated sunroof and door mirrors, and headlamp wash/wipe. Dunlop D7 tires with a larger 215 section provided better handling and feel.

For distinction, Daimler models didn't get alloy wheels as standard, but did get electrically controlled height adjusted seats, improved carpets, rear seat headrests, and smoother leather facings. For Vanden Plas models (both engine sizes) standards were further enhanced with even plusher seating, cruise control, and further boxwood inlaying.

The six-cylinder engines received a V-12-sized cooling fan that was thermostatically controlled and an oil cooler.

Automatic transmission cars got a change in the positioning of the selector, removing the detent to allow smoother operation between D and the lower 1 and 2 positions. For the six-cylinder cars the later BorgWarner Model 66 gear was fitted.

For 1982, leather trim and an integrated radio/cassette system were standard on 4.2s, while 3.4s were upgraded with electric windows and central locking as standard.

1983–1992: Long Lasting

By the end of 1982, the Daimler name would be dropped from the European market and replaced by Jaguar Sovereign, a former Daimler model. The same approach was later adopted for the UK.

All models were updated with a new center-console area incorporating a switch panel and storage tray and wood veneer. The analog clock (or onboard computer where now fitted) were moved to the upper panel and appropriate Jaguar or Daimler badging

(ABOVE) A Jaguar XJ12 Sovereign with Pepperpot alloy wheels, which continued in production until the end of 1991.

(RIGHT) Revised better quality badging for the later models.

(ABOVE) The later 4.2-liter six-cylinder XK engine equipped with fuel injection, fitted to Series 3 XJ6s.

(LEFT) The end of the line as the very last Series 3 XJ (a Daimler Double Six) left production at the end of 1992.

added to the lower panel. A further new panel carried the radio and air conditioning controls on the console, finished in black. There was also a new thicker-rimmed steering wheel.

Along with restyling of external badges, a new "Pepperpot" alloy wheels were fitted to Sovereign V-12 models. A single coachline was now applied to 3.4-liter models as were halogen headlights.

In 1984, the Sovereign XJ6 and HE (V-12) became available in the UK as the top-of-the-range Jaguar-badged models incorporating all the usual features and more. It was £1,000 cheaper than the equivalent Daimler Vanden Plas (or Jaguar Vanden Plas as it was called in the United States). The "VDP" designation continued in the United States but was dropped for the UK market. Daimler models continued elsewhere in just two variants: 4.2 and Double Six with all the features of the old VDP except for a vinyl roof covering.

There were other enhancements from 1986 on including a revised and enhanced air-conditioning system, ABS brakes, and catalytic converters, with the cars remaining relatively competitive in the marketplace despite the design's age.

Although Jaguar introduced its all new replacement XJ in 1986, the Series 3 XJ6 continued in production until April 1987. The Jaguar Sovereign HE continued until November 1991 and the Daimler Double Six until December 1992. It was an incredible run, resulting in over 168,000 Series 3 cars and an astonishing 395,000 total Series 1, 2, and 3 models.

A top-of-the-range Daimler and Jaguar Sovereign model with the on-board computer. Controlled via the illuminated push-buttons, read-outs could be obtained for distance travelled and fuel consumption.

SPECIFICATIONS

MODEL	2.8-liter Series 1	3.4-liter Series 2/3	4.2-liter Series 1	4.2-liter Series 2	4.2-liter Series 3
ENGINE SIZE	2,791cc	3,442cc	4,235cc	4,235CC	4,235cc
CARBURETION	2 x SU	2 x SU	2 x SU	2 x SU	Fuel Injection
MAXIMUM BHP	180@6,000	161@5,000	245@5,500	170@4,500	200@5,000
MAXIMUM TORQUE	182@3,750	189@3,500	283@3,750	231@3,500	236@3,750
GEARBOX	4-speed	4-speed	4-speed	4-speed	4-speed
AUTOMATIC	BW 3-speed	BW 3-speed	BW 3-speed	BW 3-speed	BW 3-speed
0 TO 60 MPH	11 sec.	10.9 sec.	8.8 sec.	9 sec.	10.5 sec.
STANDING ¼ MILE	18.1 sec.	18 sec.	16.5 sec.	17 sec.	17.6 sec.
TOP SPEED	117 mph	117 mph	124 mph	124 mph	128 mph
AVERAGE FUEL CONSUMPTION	18 mpg	17 mpg	15 mpg	15 mpg	16 mpg

MODEL	5.3-liter Series 1	5.3-liter HE Series 2/3
ENGINE SIZE	5,343cc	5,343cc
CARBURETION	4 x Stromberg	Fuel Injection
MAXIMUM BHP	253@6,000	285@5,750
MAXIMUM TORQUE	302@3,500	294@3,500
GEARBOX	n/a	n/a
AUTOMATIC	BW 3-speed	GM 3-speed
0 TO 60 MPH	7.4 sec.	7.8 sec.
STANDING ¼ MILE	15.7 sec.	15.7 sec.
TOP SPEED	150 mph	147 mph
AVERAGE FUEL CONSUMPTION	11 mpg	16 mpg

CHAPTER EIGHT

GRAND TOURER: XJ-S

A Cat of Nine Lives

The XJ-S had a lot going for it, with high hopes of challenging its major rival, the Mercedes SL. Yet mired as it was in the dismal British Leyland era, it fell into near oblivion within a few years. Fortunately, with the support of Jaguar's new boss John Egan, the XJ-S received a new lease on life. Time after time, the model was improved and modernized over an incredible 21-year lifespan to become Jaguar's most successful sporting model of all time

1975: A Major Contrast

The XJ-S derived from the XJ6 when the latter was only about a year old. Utilizing as many common components as possible would keep costs down and accelerate the new model's launch. The established excellent ride from the XJ could be maintained, while providing sporting drivers with more room inside than in the outgoing E-Type.

The XJ's floorpan was shortened 6 inches by moving the rear suspension area and bulkhead forward and stiffening up the frontal areas. The internal structure around the boot was different due to US legislation that required fuel tanks to be repositioned. To accommodate a

single fuel tank over the rear axle and behind the rear bulkhead, a platform was created. Also, to gain sufficient space for a reasonable luggage area, the boot floor was lowered and the spare wheel mounting was upright against the fuel tank.

Aerodynamicist Malcolm Sayer laid down the body design's basic principles with approval from Lyons. Unfortunately, Sayer died before the project was brought to production. Sir William retired, also before launch, but retained "consultancy control" over the final details.

The styling was low set with fewer curves and flatter paneling, presenting a more modern look with better

aerodynamics. The design had a drag coefficient of 0.772 compared to the E-Type's 0.810. One of Sayer's principal features was at the rear, where what became known as the "flying buttresses" swept down from the roof to the rear edge of the wings. The feature added style, high-speed stability, and enhanced structural integrity.

To meet federal requirements the car was equipped with 5-mile-per-hour impact-absorbing bumper bars covered in solid black plastic front and rear. Instead of different bumper treatments for various markets, the aim was to have one type for all.

The XJ-S as it was launched to the world—a culture shock for many Jaguar traditionalists with hardly any chrome, no leaping mascots, and no wood veneer inside the car. And those "flying buttresses" at the rear were highly controversial.

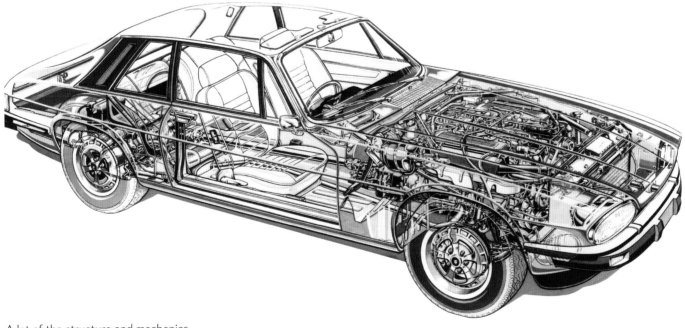

A lot of the structure and mechanics were taken from the XJ12 saloon.

The painted matte-black rear panel was short-lived on the XJ-S and looked out of place, as did the "hefty" rubber faced bumper bars.

With such a low-set frontal aspect there was no room for a traditional radiator grille, and moving away from curvature, no "mouth" like the E-Type. Instead, the nose featured a shallow flat horizontally barred grille flanked by the headlights set above the new black bumper, which contained the turn signals and space for a license plate. The body continued below bumper level incorporating a further air intake and a black rubber spoiler. The almost completely flat bonnet had no adornment at all, although the windscreen still incorporated a chromed surround. Despite the uniformity of bumper treatment for all markets, headlights could not be standardized. Two oval Cibie halogens specially designed for the car were used for the home and some overseas markets, while the US got four

The 5.3-liter V-12 in the XJ-S was the latest version of the unit with fuel injection.

The XJ-S interior was all new and totally devoid of wood but was very well equipped as standard.

A 2+2. Rear seating looked comfortable but was cramped in headroom and legroom for adults.

The unusual and unliked instrumentation for the XJ-S featured bar gauges, which were apparently more accurate than conventional analog, but did not project quality.

conventional circular tungsten lights set within chromed surrounds.

The sideview was very flat with only a hint of a swage line above wheel arch level. Window frames sloped inward towards the roof with a matte black painted finish. A quarterlight in the door did not open. Aft of the rear quarter windows, a black panel included an air extraction ventilator.

A quite shallow rear screen was enclosed within the "buttresses." The narrow flat boot lid opening incorporated a matte-black painted panel for the registration plate and badging. Twin exhaust pipes exited through the rear valance.

If the XJ-S body was a total contrast to anything else from the Jaguar stable, the interior was just as controversial—luxurious yes, but in no way traditional. Out with wood, in came plastic and black. No trim was taken from the old sports car and only switchgear and steering wheel, now leather covered, came from the XJ saloon.

There was new leather seating, fully reclining at the front with a split-type bench at the rear. Although overall there was more space than in the E-Type, rear legroom and headroom were restricted, making this 2+2 more suitable for children than adults.

The dashboard was new, a vacuum-formed black plastic type with the center console reminiscent of the XJ Series 2. Although incorporating the same heating/air conditioning system as the saloons, the instrument layout was totally new. The speedometer and rev counter with very modern legends sat aside bar type auxiliary gauges with the warning light cluster above, all unfortunately set within a quite flimsy cowling.

Pitched as one of the top-of-the-range Jaguars, the XJ-S had most features as standard equipment, like electric windows, central locking, air conditioning, and a quality sound system.

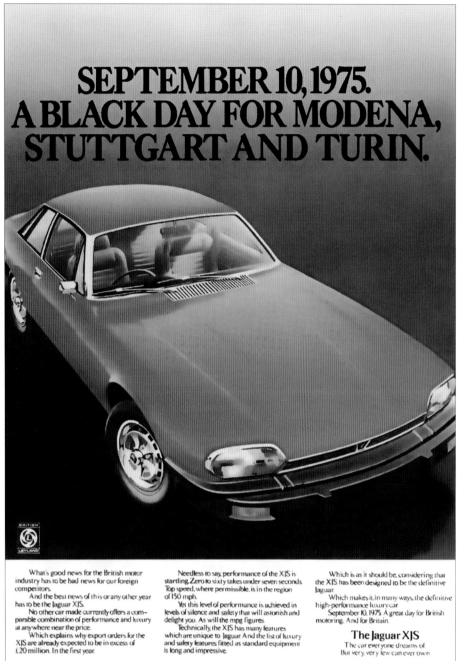

Jaguar's early promotion of the XJ-S clearly identified the market the company was aiming for with the new car.

The XJ-S was only ever destined to take Jaguar's V-12 engine and in most aspects all the mechanicals came from the saloons. The engine fitted was the latest version of the fuel injected V-12, but tougher emissions equipment for the US market cut brake horsepower there by about 40. The engine was mated either to a BorgWarner Model 12 automatic transmission or the previous E-Type's four-speed all synchromesh manual gearbox, modified for extra torque. At the rear, a Salisbury Powr-Lok differential with a 3.07 ratio was used.

Toward the ground, spring rates were changed, as the XJ-S was around 2 hundredweight lighter than the equivalent saloon, and there was a unique rear anti-roll bar. The braking system was pure XJ12. Tires were new Dunlop Formula 70 SP Super Sports on 6-inch wide GKN Kent alloy wheels.

Launched in September 1975, the XJ-S was one of the most expensive Jaguar/Daimler models one could buy at £8,900, but significantly cheaper than the equivalent Mercedes.

1976–1980: Refining the Package

The XJ-S didn't require much refinement to improve the package, and some changes followed the saloons. In April 1977, the GM 400 automatic transmission replaced the BorgWarner unit. In 1979 the manual transmission was dropped.

From 1977 there were minor trim changes. The radiator grille's black trim was replaced by chrome, and the matte-black rear boot panel was dropped for a body-color finish. Internally, the instruments lost their silvered bezels, better quality carpeting was fitted, and twin color console finishes were introduced.

In 1980, amidst complaints about the poor fuel economy, Jaguar introduced the new Lucas/Bosch digital electronic fuel injection. This not only marginally improved fuel consumption but also provided a boost in power. That change, along with a three-way converter, increased output by 18 brake horsepower in the US.

Despite these changes, sales of the XJ-S were disappointing. In its first full year of production, 1976, only 3,082 buyers came forth (compared to 7,847 for the first full year of E-Type Series 3). Poor sales dropped production to a mere 1,057 by 1980. Something had to be done if the car was to survive.

1981–1982: Rebirth

Jaguar's board of directors discussed axing the XJ-S. The new front man, John Egan, was certain the car could be improved. The three elements of concern were reliability/quality, fuel economy, and "Jaguarness."

Amidst industrial unrest, a lack of profitability, and poor sales, Egan encouraged the workforce, installed new parts-testing procedures, and impressed on suppliers the importance of improving their standards in order to revitalize the company's cars.

An improved cylinder head by Michael May (discussed more fully in Chapter 8) was released in July 1981. This change alone improved fuel economy, and when combined with a new

rear axle ratio of 2.88:1, increased the car's top speed potential to 155 miles per hour. Sadly, emissions controls in the States negated the improvements.

Some cosmetic improvements were made to emphasize the positive changes to the XJ-S and help revitalize sales. Externally, a circular jaguar-head badge appeared in the center of the bonnet. Bumpers from the XJ Series 3 featured chrome trim. Side turn signals were added to the front wings (previously a standard feature on US cars), and a twin coachline was applied along the swage line. New unique "Starfish" alloy wheels were fitted with 215 section Dunlop D7 tires. The scuttle-mounted air intake and wiper arms were now finished in black.

The HE version was a godsend to the car's fortunes. A little chrome, tidying up the bumper arrangement, adding a coachline to break up the bland side of the car, plus major improvements to engine efficiency effectively relaunched the XJ-S.

The "new" model was called the XJ-S HE (for High Efficiency) with an appropriate "HE" badge adorning the boot lid.

Internally, burr elm wood veneer trim was added to the dashboard and doors, the latter with leather trim rather than vinyl. An XJ Series 3 steering wheel with leather rim was adopted. Feature changes included improved central locking, courtesy light delay, and door open lighting in the doors.

The HE saw a return to wood veneer and a new style of steering wheel.

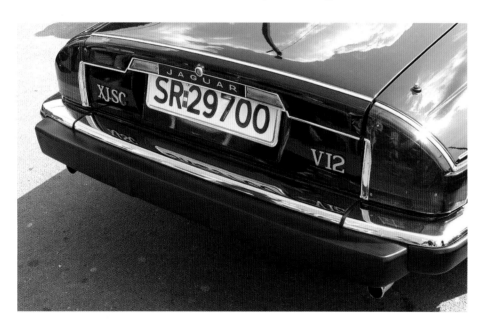

Although it was always planned that the bumper arrangements on the XJ-S would satisfy all market needs, legislation changes affected some markets, making such rubber additions necessary.

Finally, with improved build quality, the service interval for the US market was increased to 24 months/36,000 miles.

At launch in the UK, the price for the HE was £18,950, which was actually 4 percent cheaper than the old pre-HE model! All these changes were enough to revitalize the market and sales improved.

1983–1984: Expanding the Range

To further expand market share, Egan called for a new engine and gearbox. Jaguar had historically chosen a sporting model to launch its new engines (the XK six-cylinder twin-cam unit in the XK120 in 1948, and the V-12 in the E-Type in 1971). So, it was to be the XJ-S to receive the new multivalve six-cylinder AJ6 engine, finally destined to replace the aging XK unit in all cars.

The new engine was all aluminum, retaining the twin overhead camshaft arrangement operated by chains, but in a four-valves-per cylinder head. A seven main bearing cast iron crankshaft was fitted in the block with interference-fit dry liners for the pistons. The new engine had dimensions of 91×92 millimeters giving a capacity of 3,590cc (3.6 liters).

Fitted with the Lucas/Bosch P-digital fuel injection system with a 9.6:1 compression ratio, it provided 225 brake horsepower at 5,300 rpm and 240 pounds of torque at 4,000 rpm, all of which compared favorably with Jaguar's other engines. There was also a weight savings of 430 pounds—123 pounds lighter than the XK and 210 pounds lighter than the V-12.

To fit in the bay, the engine was canted over 15 degrees to the exhaust side. Spring rates were altered and the rear anti-roll bar was deleted.

Initially the 3.6-liter was available only with a manual transmission, a new five-speed Getrag, shared with many

(ABOVE) The XJ-SC was a stop-gap model until Jaguar could produce a full convertible. Despite the manual nature of the top, it provided open-top motoring to XJ-S owners for the first time. The unusual design was free of wind buffeting with the top down, even at high speeds.

(RIGHT) Jaguar's AJ6 multivalve six-cylinder engine, destined to replace the XK, which had never appeared in the XJ-S.

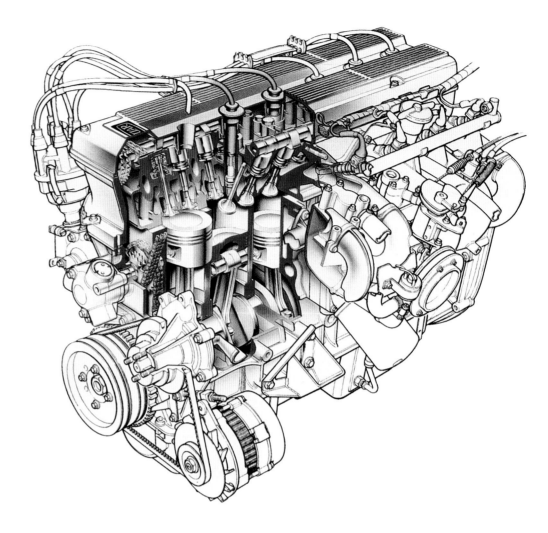

BMW models. This provided a sportier feel to the car and was mated to a 3.54:1 rear axle ratio, but it was never sold in the States.

Stylistically the 3.6-liter model had a new bonnet with a raised hump, Pepperpot alloy wheels taken from the XJ Sovereign and "3.6" badging on the boot lid. Internally it was much as before, although half leather trim was the norm with cloth inserts; full leather could be ordered.

Coinciding with the 3.6-liter XJ-S coupe, there was another new model launched with that engine, the XJ-SC (Cabriolet), a design midpoint between a closed coupe and a full convertible. The Cabriolet was conceived as a cost-effective way to reintroduce an open model, missing since the demise of the E-Type.

The cabriolet design retained the door and rear-quarter window frames but replaced the metal roof with removable components. Across the center of the car, a hoop connected to each B post to preserve rigidity. Behind it, a lined top could be folded down to rest on the rear deck. Ahead of it, two removable fiberglass

THE MAKING OF THE CABRIOLET

The XJ-SC initially went through a convoluted process of manufacture, starting with a conventional coupe bodyshell built at Jaguar's Castle Bromwich body plant. With the roof and rear header panel omitted, the bodies were shipped to the Park Sheet Metal Company in Coventry, where they removed the "buttresses" and welded in a new rear deck. They also strengthened the floorpan around the transmission tunnel and fitted a new crossmember subframe around the rear of the car for extra rigidity.

The bodyshells were then returned to the Castle Bromwich facility for painting, followed by transfer to Jaguar's Browns Lane Assembly Plant for the addition of the mechanical parts and trim.

From there, the cars were moved to the Aston Martin Tickford facility in Nuneaton for installation of the Targa panels and fold-down top, then returning for the final time to Browns Lane for detailing, road testing, and quality checking. From there they were dispatched to the appropriate dealership.

A slight change came in 1985, when the Targa arrangement was fitted in-house at Browns Lane.

A US CONVERTIBLE

The convertible by Hess & Eisenhardt, with whom Jaguar contracted to support the US market until their own convertible was ready for production.

Jaguar's fear of a US ban on convertibles never materialized and the company decided to jump on what it saw as an opportunity. A factory-built convertible was back on the agenda. In the interim, an 18-month contract was set up with a coach-building firm to assemble convertibles in the States.

In October 1986, Hess & Eisenhardt in Ohio offered XJ-S coupes modified to convertibles by special order, supplied through Jaguar's own dealerships. On receiving UK-built HE Coupes, H&E stripped out the interior, welded strengthening members to the sills, then removed the roofs and rear quarter panels. The roof was cut short of the windscreen to allow room for the top's securing catches. New paneling was welded in accordingly, and painting to match the original was completed. The existing fuel tank had to be replaced by two smaller ones. The specially made top was electrically operated and designed to fold down into the cavity created in the rear compartment/fuel tank area, and a trimmed cover was made to fit over it. Door window frames were eliminated so the glass formed a seal with the hood.

The mechanical aspects remained unchanged. The conversion was carried out on new cars only, so when returned to the dealerships they were still eligible for the full Jaguar warranty. Approximately 2,000 H&E convertibles were produced.

Targa panels trimmed to match the top covered front passengers. A matching fabric tonneau cover was trimmed for the top when folded down. Jaguar also produced a double-skinned fiberglass hardtop available at extra cost, fabric covered and with a heated rear screen. The Cabriolet didn't have rear seating because it was a safety issue, although some owners subsequently had them fitted from a Coupe.

The two 3.6-liter XJ-Ss were aimed at a different market, satisfying a wider demand to own a Jaguar sporting model, while the HE continued to sell exceptionally well. In 1984, Jaguar sold a record 6,028 XJ-Ss, half of them in the States.

During this period, the 3.6-liter XJ-SC was equipped with an on-board computer, cruise control, headlamp wash/wipe, and a better audio system, all previously extra-cost items.

1985–1986: More Expansion

The V-12 continued to sell in larger numbers, so in July 1985 Jaguar announced another new model, the XJ-SC HE. Now fitted with the more powerful engine, and a limited slip differential, the extra Cabriolet offered the same appointments as the 3.6.

Coinciding with the launch of the XJ-SC HE, the HE coupe's badging was changed to V-12 XJ-S and it received walnut veneer to replace the elm previously used.

The XJ-S range went from strength to strength with another record of 9,052 being produced to meet demand in 1986.

1987: Further Enhancements

In February, the 3.6-liter models were made available with an automatic transmission for the first time, a new ZF 4HP 22 four-speed unit. This was a three-speed epicycle gearbox with the third gear being the direct drive, and the fourth more of an overdrive ratio for optimum economy.

As the 3.6-liter engine was now in full production mode in Jaguar's new XJ saloon, many changes took place. These included a fully electronic ignition system providing better economy.

All XJ-S models benefited from trim changes including "Jaguar" etched, bright treadplates, heated door mirrors and washer jets, a revised center console with new switchgear, and even a new steering wheel design. Heated seats with electrical lumbar support were standard on V-12s, optional on 3.6s, as was auxiliary lighting at the front.

To differentiate the six-cylinder cars from the 12s, in September Jaguar introduced the Sports Handling Pack for the former. Front and rear spring rates were increased with uprated Boge shock absorbers, the rear anti-roll bar was reinstated, and the front anti-roll bar was strengthened. Power assistance on the steering was reduced with stiffer rack mounting bushings, along with wider section 235/60 VR Pirelli P600 tires.

The 3.6-liter Cabriolet model was discontinued in September and just after the end of the year the V-12 was also discontinued to make way for another new model. Another record (9,714) XJ-Ss were delivered in 1987.

1988: Topless at Last!

Jaguar's Cabriolet had satisfied only a niche area of the market for convertibles and with improved sales and more money available, the company set about designing its own two-seater XJ-S. The German Karmann company was hired to engineer the work and build the first fully working prototypes. Jaguar's specifications were for a stylistically pleasing top folded or in the erect position, fully electrically operated, and with a finished car retaining all the refinement of the hardtop models.

Jaguar's own full convertible, the premium priced, luxury model in the range was very well received. Introduced in 1988 and initially only available with the V-12 engine.

The luxuriously appointed interior of the XJ-S V-12 convertible with walnut veneer, best quality leather upholstery, and most options as standard.

JAGUARSPORT

The JaguarSport venture produced the XJR-S, here seen in 6.0-liter form with the full body kit and Speedline alloy wheels (set against the racetrack backdrop and Jaguar's Le Mans–winning car).

A joint venture set up by Jaguar and TWR created JaguarSport in 1988 to produce a modified XJ-S called the XJR-S, to be sold through a select number of Jaguar dealerships. The idea was to develop a younger market for the XJ-S.

The cars underwent assembly at Browns Lane as a normal car but were then shipped down to TWR's facility near Oxford for conversion and finishing. Special body kits and Speedline alloy wheels with lower profile Pirelli tires were fitted, and minor changes to the interior took place. Initially the XJR-S cars were mechanically similar to factory models, but later a 6.0-liter V-12 engine was developed before Jaguar's own engine in 1993. The cars were equipped with Zetec sequential ignition and modified intakes, which along with altered spring rates and the new wheels/tires boosted performance.

When the facelifted XJS was announced, JaguarSport versions were also produced. Most were coupes, but some convertibles were made for the US market. In the 1990s, a JaguarSport cost over £6,000 more than a production car, and after around 800 were produced the project was dropped in 1993.

Approximately a third of the body's panels were changed or amended to strengthen the bodyshell, which included fitting steel tubes within the sills and A posts. Bodies were assembled at Jaguar's Castle Bromwich body plant using a new £3.6 million system. Precision building was necessary to ensure a perfect fit of the top to the body. The frames were trimmed at Jaguar's own Browns Lane Assembly Plant.

The XJ-S convertible was the first soft-top in the world to be fitted with a conventional heated glass rear screen. To eliminate the need for part of the roof to be retained above the windscreen surround, the top's forward securing catches were fitted to the top of each A post. The top complete with the rear quarter windows could be electrically retracted from a switch on the center console in just 12 seconds.

The rear seats were removed, replaced with a lockable box area similar to the Cabriolet's, behind which were stored the hood motor and hydraulics.

Although the suspension remained unchanged from the standard V-12 coupe, new Lattice style alloy wheels were fitted, which were adopted throughout the range thereafter. Earlier that year, a new Teves braking system with ABS had been fitted to all models, and this was carried over to the convertible.

The convertible was the most expensive of all XJ-S models and was therefore equipped luxuriously with no added extras. Initially it cost £36,000 in the UK and soon demand was so high that exorbitant prices were being quoted by those who saw the opportunity to "make a packet" from the sale of their new car!

1989–1990: Special Editions and More

For the US market in 1989 a "Collection Rouge" XJ-S V-12 was announced. This was a limited-production model painted

The US Collection Rouge model.

One of the XJ-S special editions was the Le Mans to celebrate the 1988 and 1990 Le Mans wins. The car even had its own quality brochure.

in Signal Red with a gold coachline and diamond polished red-spoked Lattice alloy wheels, plus unique badging. Internally, the magnolia leather trim was piped in red, with magnolia leather steering wheel and gear knob and burr elm wood veneer. Costs were $51,000 for the coupe and $57,000 for the convertible.

For 1990, there were more changes to the XJ-S range. The V-12 engine received a new Marelli digital ignition system and catalytic converters in most markets. The Sports Pack from the 3.6 liter could now be specified for V-12 coupes. Internally, there was another new steering wheel,

revised stalks on the column, and a new column-mounted ignition switch.

More special editions came along in 1990, the first for the home market being the Le Mans XJ-S V-12. Based on the standard V-12 coupe and limited to 200 examples, the model was built to commemorate Jaguar's Le Mans success

CELEBRITY STATUS!

Who was the biggest celebrity ever to own an XJS?—Barbie! The Mattel toy company's doll is an icon of the toy world. Over the years, she has owned several cars, but the Jaguar XJS has to have been her favorite.

When Mattel asked for the rights to use the model and name, who would have refused? Later, when asked to prepare a real car for promotional purposes, Jaguar resurrected an ex-press photography convertible that had been lying around the factory due to water damage during a photo-shoot abroad. The car was repainted in Barbie's special pink and used on numerous occasions to promote Barbie—and Jaguar.

---

that year. It featured the four-headlight treatment, 16-inch Lattice alloys, Sports suspension, Autolux full leather trim on redesigned seating, high-contrast walnut veneer, and Wilton carpeting. Sill treadplates were etched with "Le Mans V-12" and the car's edition number.

For the US market, they had the "Classic Collection" model with unique paint and trim offerings along with gold plated badges.

Sales figures were still buoyant for this now aged model. In 1989, 11,207 cars were turned out. The figure dropped to 9,226 in 1990. Time for a rethink!

1991–1992: Fundamental Facelift

Although the XJ-S had been selling well, it was long overdue for a facelift. Ford took over Jaguar in 1989, and their first involvement was to sanction improvements to this important model, to take effect from April. Around £50 million was invested in changes to styling, trim, and the engine. The car was also renamed, now simply "XJ," not "XJ-S."

Starting with the body, out of the 490 panels it took to build an XJ-S, 180 of them were revised. The alterations cut

down costs and production time. For example, the new rear wing was made from one pressing, whereas the old car had five.

Up front came a black grille, chromed horizontal strip, and different oval lighting, which was now fitted to US cars. The 3.6-liter hump bonnet was standardized on all models. From the side, new door pressings eliminated the window frames and the quarterlights. The roofline was slightly flattened with a marginally changed rake to the rear screen, and although there was no change to the inner paneling around the rear side windows, external revisions to the glass and trim area gave the appearance of a wider area. The biggest change came at the rear with a host of new paneling and trim, neutral density lighting and badging.

Internally, there were also lots of changes. First, although the old dashboard layout looked similar, the top roll was now available in a choice of colors. The old instrument pack with bar instruments was thrown out, replaced with the same instrument binnacle as used in the XJ40 saloon range, containing clearly marked and analog gauges. There was new switchgear, the main lighting switch moved to the steering column, a new style on-board computer, and the adoption of the XJ40 saloon's air conditioning system and layout. There was new seating front and rear, the front electrically operated on all models. The XJ40 audio system was also fitted but with a new face-plate. The boot interior was better finished, now with full carpeting of a better quality.

Mechanically, the V-12 models got another new digital fuel control system with an on-board diagnostic system to help dealers identify issues. The engine bay layout was improved. The six-cylinder models received the relatively new 4.0-liter engine first fitted to the XJ40 saloon in 1990 (see Chapter

(TOP) Comparison of the 1991 facelift coupe and the earlier XJ-S, showing the restyled rear end, the most easily identifiable feature of the new car.

(BOTTOM) The clever use of glass and trim hides the fact that the actual window area is no bigger than previously!

(TOP) A facelift 4.0-liter convertible.

(BOTTOM) The analog instrument pack fitted from the facelift models was welcomed by all.

In the 1990s, Jaguar introduced an Insignia project. Using their in-house Special Vehicle Operations Department, they offered a bespoke range of finishes, trim, and color schemes for the XJS (and XJ40 saloons).

10). Fitted with a new catalytic converter, the XJS could be sold in more countries. The new engine produced an extra 6 percent in maximum power and 14 percent extra torque.

To accompany the new engine, there were two new transmissions. The Getrag 290 was an upgraded version of the existing five-speed manual gearbox, with a three-plane gate for better gear selection, and the lever located slightly further back to improve accessibility. A larger 11-inch clutch and a new

twin-mass flywheel completed the "stick shift" change.

The new ZF 4HP 24E automatic provided full electronic control, the module continually monitoring the speed of the output shaft and throttle position to interpret the ideal gearing.

By the end of 1991, all pre-facelift models had been discontinued after total production of 88,159 cars.

It wasn't until 1992 that the 4.0-liter engine could be specified in a convertible, but in automatic transmission

form. In the same year, Jaguar started to fit a stainless steel cruciform under the front of all convertibles for greater strength. The steering column and wheel were adapted to take a driver's air bag, surprisingly an extra-cost option at the time!

By the end of the year, Jaguar had turned out 7,361 facelift models, to some extent stemming the decline in sales but still not matching earlier successes.

(ABOVE) The final styling changes to the XJS involved new preformed body color bumper bars making the car look still modern. By this time, quality was at its best.

1993: Yet More Changes and Enhancements

The year started with more modifications to the XJS range. Finally, after the announcement of the 4.0-liter convertible, it became available with manual transmission. Although the Sports Pack was still a standard feature on the smaller-engine cars, one could be ordered with the softer "touring" suspension.

To improve rigidity, rear struts were added to convertibles.

Internally, a new steering column released an extra 2 inches for the driver and the driver's-side air bag was standardized. The seats were given extended sides to satisfy the needs of taller drivers and front seat passengers, and velour upholstery replaced the cloth type previously used.

Due to changes in tax regulations, V-12 cars were no longer imported in the US, so the range of XJSs was reduced to just the 4.0-liter coupe and convertible, both with automatic transmission. However, manual transmission could still be supplied to special order. Prices had increased by this time with the coupe

costing $49,750 and the convertible $56,750.

Still being sold in other countries, the V-12 underwent further changes. In May 1993, a larger 6.0-liter version was announced. The capacity increase to 5,994cc came about by increasing the stroke from 70 to 78.5 millimeters with an unchanged bore. There was a new cylinder head with flat-top piston design, a reduced compression ratio, new cylinder liners, revised intake valves, new camshaft profile and a new forged crankshaft—also a new torque converter, low-loss catalyst system, in-tank fuel pump system, even a new starter and alternator, plus a Marelli engine management system. Maximum power increased

The last version of the V-12 engine to be fitted to XJSs, the 6.0-liter, 318-brake-horsepower version.

The new rear bumper treatment blended in well with the XJS styling.

to 308 brake horsepower at 5,350 rpm and torque of 355 pounds at 2,850 rpm. Under the hood, a smaller air conditioning compressor and a well-designed cover plate for a lot of the pipework and wiring on top of the engine tidied up the appearance.

A new GM 4L80-E four-speed electronic automatic transmission used the same gear train as the old GM400 box but with an extra set of gears to provide a fourth overdrive ratio. It also had a lock-up clutch for added economy. The new gearbox incorporated both a Sport and Normal modes operable from the center console switch. The transmission was mated to a control module communicating electronically with the Marelli engine-management system to reduce engine torque during shifts. The system also included a self-diagnostic facility recording malfunctions.

The new engine/gearbox combination improved top speed by 13 miles per hour and reduced the 0- to 60-miles-per-hour time by over a second.

A new form of Sports suspension package was fitted to all XJSs at this time with Bilstein shock absorbers, revised spring rates, and front anti-roll bar; yet again the rear anti-roll bar was deleted! Cars could be specified with low-profile Pirelli P600 tires on new 16-inch wheels.

A big change in the braking system saw the rear discs (with new calipers) moved outboard for the first time since IRS was introduced back in 1961! Steering was changed with a new ZF rack.

For the first time in a convertible, a 2+2 seating arrangement came about with a complete redesign of the rear bodyshell. A complete subassembly was welded on top of the existing subframe, forming the rear seat pan and back, replacing the existing storage box, and making space for occasional seating. This had the added bonus of improving structural integrity. By making the rear screen shallower and moving the roof pump to the boot, this left space for the hood to fold down in the same way as previously. A two-seater could still be ordered using the earlier bodyshell.

All XJSs benefitted from styling changes. The old-style rubber faced bumper bars gave way to all enveloping plastic finished to body color. At the front, the bumper incorporated a new spoiler design and under-bumper grille.

At the rear, rectangular tailpipes exited from the new style bumper.

1994–1996: The End Approaches

Jaguar became one of the first luxury car manufacturers to fit their cars with passenger air bags, which the XJS received in 1994. The air bag replaced the glove box with changes to the under-dashboard trim. The AJ6 engine got a new cam cover incorporating the oil filler and an integrated crankcase breather, and there was a one-piece cast intake manifold.

Just after this in the summer of 1994, the XJS received a new AJ16 4.0-liter six-cylinder engine (see Chapter 10). Along with other innovations, the design incorporated the new concept of individual on-plug coils. Horsepower increased to 238 with an extra 4 pounds of torque, along with critical improved fuel economy and emissions performance.

An all new engine-management system, fully programmable, communicated with the automatic transmission. There was also a new immobilizer device using a coded transponder in the key fob.

More changes followed in 1994 in styling and trim. To differentiate the 6- and 12-cylinder models a degree of color coordination in exterior trim was used on the 4.0-liter models along with new alloy wheel combinations. The V-12s got a new twin coachline along the side and a matte black radiator grille. Yet another style of seating was designed with integral headrests, and there were front seat pockets built in for extra storage. The 4.0-liter models got cloth center panels and shallow-grain leather outside seat facings, while the V-12s got the finest Autolux leather throughout with contrast piping. New carpeting with contrast piping and bright alloy treadplates followed for all models. Finally, a new,

more powerful audio system was fitted with a detachable front panel.

In 1995, Jaguar's 60th anniversary and 20 years of XJS production provided an ideal opportunity to raise the profile of the now aging model. It was commonly known that the model was going to be replaced shortly, so to boost sales leading up to that the 4.0-liter Celebration special edition was created.

Chrome finish instead of color coordination returned to exterior trim, a reversion to a black radiator grille and coachlines, plus extra chrome trim at the rear. Diamond turned Aerosport alloy wheels, available in five-spoke chrome in the US, were accompanied by gold-enameled bonnet badging.

For the interior, the Celebration took on all the luxuries of the V-12 with Sapwood veneer and a half wood/leather steering wheel.

By 1995, V-12-engined XJSs had virtually been abandoned, only built to special order. Production of the other models continued until April of 1996 when the very last two cars left the line, a 4.0-liter convertible and a 6.0-liter coupe, both destined for Jaguar's own Heritage collection.

A grand total of 115,413 XKSs were built over a 21-year period, a phenomenal success for Jaguar, but it was time for a "new kid on the block."

The later AJ-16 engine fitted to the last of the XJSs.

(ABOVE) The final models, the Celebration 4.0-liter coupe and convertible with diamond turned alloy wheels and fully fitted interiors.

(LEFT) Later interior of the XJS with new seating, steering wheel, center console, and controls. These last, exceptionally well-equipped cars were the best models produced.

XJS RACING EXPLOITS

The US-based Group 44 racing team under Bob Tullius convinced Jaguar to allow them to enter the XJ-S in the 1976 Sport Car Club of America's TransAm series. The first and only race win that year came on the second outing at Lime Rock. The following year brought ten races and five wins. In 1978, Jaguar won the Manufacturers' Championship.

The XJ-S returned to US racing in 1981. A highly modified car developing 525 brake horsepower captured a total of 29 wins.

In 1981, Tom Walkinshaw of TWR Racing in the UK approached Jaguar to allow cars to enter the European Touring Car Championship to compete alongside arch rivals BMW. For the 1982 season, a couple of cars were entered without direct involvement of Jaguar and won four races. For 1983, a more concerted effort with Jaguar's support achieved five wins.

For 1984, three cars were entered achieving seven wins and the Championship.

Twenty-one years of continual production, from the 1970s XJ-S (right), 1980s HE (center), and 1990 last-off-the-line XJS, Jaguar's most successful sports car design ever.

SPECIFICATIONS

MODEL	5.3-liter XJ-S	5.3-liter XJ-S HE	3.6-liter XJ-S	5.3-liter XJ-S	4.0-liter XJ-S
ENGINE SIZE	5,343cc	5,343cc	3,590cc	5,343cc	3,980cc
CARBURETION	Fuel Injection	Fuel Injection	Fuel Injection	Fuel Injection	Fuel Injection
MAXIMUM BHP	285@5,500	295@5,000	225@5,300	290@5,750	223@4,750
MAXIMUM TORQUE	294@3,500	318@3,500	240@4,000	309@3,150	277@3,650
GEARBOX	4-speed	n/a	5-speed	n/a	5-speed
AUTOMATIC	3-speed	3-speed	4-speed	3-speed	4-speed
0 TO 60 MPH	7.5 sec.	7.5 sec.	7.8 sec.	7.5 sec.	8.7 sec.
STANDING ¼ MILE	15.7 sec.	15.6 sec.	16 sec.	15 sec.	16.7 sec.
TOP SPEED	142 mph	153 mph	134 mph	153 mph	136 mph
AVERAGE FUEL CONSUMPTION	14 mpg	16 mpg	18 mpg	17 mpg	18.5 mpg

MODEL	6.0-liter XJ-S
ENGINE SIZE	5,994cc
CARBURETION	Fuel Injection
MAXIMUM BHP	308@5,350
MAXIMUM TORQUE	355@2,850
GEARBOX	n/a
AUTOMATIC	4-speed
0 TO 60 MPH	8 sec.
STANDING ¼ MILE	16.3 sec.
TOP SPEED	150 mph
AVERAGE FUEL CONSUMPTION	15 mpg

CHAPTER NINE

BIG SALOON DEVELOPMENT

XJ40 to X-308

With the XJ name becoming so well established as a Jaguar luxury model, it was natural to retain it for subsequent models. The first entirely new car to bear it came in 1986 (code named XJ40*). Although this car would be short lived compared to its predecessor, it formed the basis of two other later XJs, coded the X-300* (1994 to 1997) and the X-308* (1997 to 2002). These three model ranges saw Jaguar develop new manufacturing techniques and technologies, including new engines, yet they retained many of the respected traditional elements that set Jaguar apart from other brands.

1986: The Birth of an Entirely New Car

Under British Leyland, the XJ40 went through a traumatic and elongated gestation period. It wasn't until 1980 that Jaguar got the go-ahead from BL for the car. Over £70 million was invested and it became the most tried and tested car—covering over 5 million miles— that Jaguar had ever produced. The motivation for all this was to improve productivity, reliability, and quality; make the car more economical and technologically advanced; and retain the enviable position in the market place the XJ Series had gained over the years.

Under British Leyland, there was always the "threat" of Jaguar being told to abandon their own engines in favor of the Rover V-8. Because its own engines were so integral to the mystique, Jaguar's designers cleverly ensured that the inner structure at the front end would not accommodate the wide V-8 engine. This move would bite back during later efforts to fit their own V-12 engine!

With some in-house confusion over whether to retain the four-headlamp treatment of previous XJs, a compromise

To avoid confusion between models bearing the same XJ name, the Jaguar codes will be used to identify the different cars.

The XJ40 was a more angular shape following the fashion of the day. This was an early XJ6 model with conventional headlights and plastic wheel trim over steel wheels.

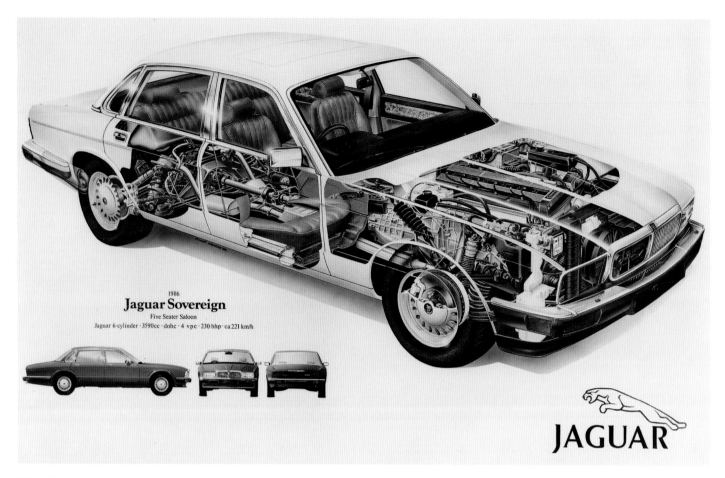

Virtually everything was new about the XJ40 from its bodyshell assembled from fewer panels, to trim, mechanicals, and electronics.

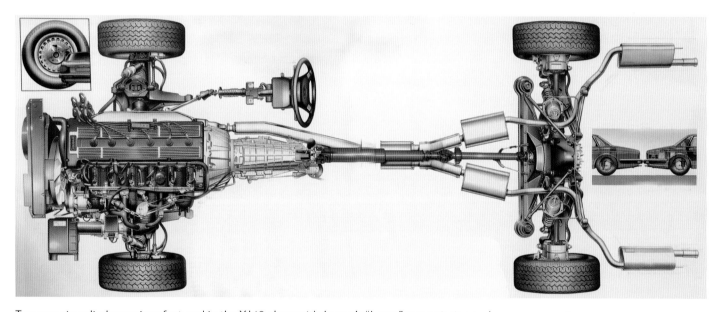

Two new six-cylinder engines featured in the XJ40 along with Jaguar's "J-gate" transmission and a new suspension.

was reached, with these and more modern rectangular singular integrated units used for the various models. New rubber bumpers were fitted with chromed top blades, now incorporating auxiliary lighting with reflectors to the sides. Under the bumper, a new air dam was designed to aid aerodynamics.

A rectangular radiator grille was retained, but it was more upright. The bonnet panel was very flat with a slightly raised central area but no chrome adornment, leading to an enlarged windscreen area, and for the first time in a Jaguar, a single centrally mounted windscreen wiper.

The relatively flat body sides incorporated a waist swage line (the Daimler model would feature full-length chrome trims), and a further horizontal line at a lower level extending the full length between wheel arches. Jaguar subtly retained the raised haunch over the rear wings. The large side window area was of conventional six-light style, without any quarterlights and retaining brightwork in various forms according to model.

Big changes in rear styling included a raised and flat boot lid which, to help aerodynamics, incorporated a very slight lip continued to the rear wings.

The whole effect of the new styling was much more angular in design, yet it retained many subtle features that made it an easily recognizable Jaguar. As to aerodynamics, the car's drag figures were 0.762, compared to the Series 3 at 0.849.

In manufacturing the new bodyshell it was essential to cut costs and improve quality, hence the number of individual pressings was reduced by 25 percent. For example, the body side, which used 20 individual welded panels on the Series 3, was just 1 on the XJ40. The body was still of monocoque construction, but it now incorporated two crush tubes running the length of the engine bay, which dissipated energy rather than sending the crash force through the passenger

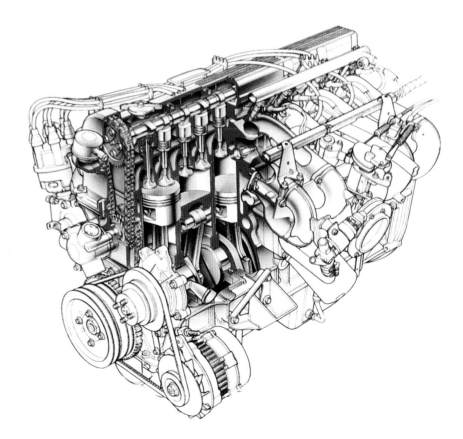

The single-use 2.9-liter engine, derived from half a V-12, was fitted only to the XJ40 model and for a limited period.

The XJ40's electronics were the most advanced of any production car when it was announced.

The interior of the early XJ40 with the unique dashboard instrument layout and J-gate transmission selector on the center console.

The interior of a base model 2.9-liter XJ6 with tweed upholstery and manual transmission. Note also the straight-grained woodwork.

compartment. Other changes to the bodywork ensured a much more rigid structure than the previous model.

Extensive use was made of zinc-coated steel, lead loading was eliminated, and a more durable, high-quality paint finish incorporated clear coat.

The AJ6 multivalve six-cylinder 3.6-liter engine first launched in the XJ-S in 1983 was available up front, with improvements including quieter valvetrain, modified cam profiles, lighter bucket tappets, new timing tensioners, and better overall engine balance.

The AJ6 engine was the direct replacement for the aged XK unit, but there was a further new engine unique to the XJ40, which stayed in production only until 1991. Of 2.9 liters capacity, it employed the same cylinder block as the 3.6-liter unit (with the same bore); the rest of the engine was similar to Jaguar's 5.3-liter V-12 design. A short stroke of 74.8 millimeters achieved a displacement of 2,919cc; a high-efficiency single overhead camshaft cylinder head allowed for a much higher compression ratio of 12.5:1. Despite its relatively small capacity, the 2.9 achieved 40

brake horsepower more than the old XK 3.4-liter six-cylinder engine. Produced extensively of aluminum, the new engines created large weight savings over the old XK unit: 25 percent for the 2.9 liter and 30 percent for the 3.6.

The strategy behind the 2.9-liter engine was to enable Jaguar to market a cheaper alternative and cater to countries where vehicle taxation was based on performance and engine size.

With the AJ6 engines, Jaguar offered the five-speed Getrag 265 manual gearbox as an option to an automatic. This option was available for the XJ40, albeit with a larger clutch master cylinder to provide more disengagement travel. The ZF 4HP 22 automatic transmission from the AJ6 equipped XJ-Ss was also used in the XJ40, but with an entirely new Jaguar-designed gear selection method that became known as the "J-Gate." The gate formed a "J" with normal gear selection taking place on the righthand side. When in "drive," the selector could be moved over to the left for selection of intermediary gears.

On the front suspension, double wishbones of unequal length and uprights in forged steel with pivoted angles provided anti-dive characteristics. The most significant difference was that the pitch control arms now faced rearwards, anchoring to a stiffer part of the body structure. An all-new subframe was fabricated from upper and lower pressings formed into a beam of simpler construction. It was filled with foam to reduce noise.

The fulcrums and mounting points were machined after assembly to ensure greater accuracy, eliminating the need for camber or steering rack height adjustment.

The engines were slanted over 15 degrees to reduce engine bay height and the engine mounts were placed so that the line between them passed through the minimum inertia axis of the engine, minimizing movement. A third engine mount connected directly to the body.

The rear suspension was also new and for the first time since Jaguar went "independent," had outboard disc brakes. The suspension embodied a pendulum arrangement allowing fore and aft movement of the lower wishbone inner fulcrum but retained a high degree of lateral stiffness; its aim was to minimize road noise while maintaining accurate geometry for best handling.

A full depth tubular driveshaft and hook joints were used as the top link of each suspension and the lower wishbone was fabricated from two pressings. Because of the outboard discs, each wishbone would take brake torque as well as power and cornering reaction forces.

Another change from the old IRS was use of only a single coil spring and shock absorber per side, the latter containing the bump and rebound stops. The inner fulcrum of the lower wishbone was angled to give both anti-brake-dive and anti-acceleration squat characteristics. The rear subframe was mounted on two rubber bushings at the front and a pair of angled links at the rear.

Another new technology for Jaguar was a self-leveling device for the rear suspension (standard on some models). An engine-driven pump provided hydraulic power for both the self-leveling and the braking system. Using struts instead of conventional shock absorbers to avoid height correction when fitted, each contained a gas accumulator that implemented the height changes with hydraulic pressure in response to an electromechanical sensor connected to the righthand rear wishbone.

Apart from periodic greasing of the rear driveshaft joints, both front and rear suspension were maintenance free.

Steering followed conventional Jaguar practice via power-assisted rack and pinion. For the braking system, there were two major changes: a brake power boost system (instead of a conventional vacuum servo) and ABS.

The power boost system took its power from the engine-driven pump serving the self-leveling device where fitted. A pressure accumulator ensured that even if the engine quit, the system

The luxurious interior of the Daimler model, also featured for some overseas markets as the Vanden Plas, featuring appointments such as individualized rear seating and occasional tables.

would store sufficient pressure to provide between 8 and 20 stops, more than with a conventional servo.

The ABS system was developed by Bosch. It incorporated a sensor on each wheel feeding information back to an anti-lock processor, which controlled the braking system via three channels—one for each front wheel and one for both the rear brakes. The system incorporated a yaw control to accommodate widely differing side-to-side braking requirements.

New metric TD-sized road wheels incorporated a bead groove for the specially developed Dunlop or Michelin 220/65 VR 390 tires. The system inhibited a punctured tire from coming off the rim and enabled the driver to carry on at low speed for a while, driving on a flat tire.

When launched, the XJ40 was heralded as the most electronically

advanced car of its day. There was a wiring system with more reliable connectors, a significant increase in the number of onboard microprocessors, and an advanced diagnostic system. Jaguar adopted a low-current, high-duty earth line switching system carrying just 5 volts instead of the usual 12. No fewer than 90 electronic relays were fitted to the car and there were over 1,700 pieces of electrical wire, all of a higher quality than before. A new Jaguar Diagnostic System was designed for the car, operable from the dealerships via a new computer with connection points and probes for ease of fault finding.

For the interior, the parameters were less weight, better quality, modernity, and that unique Jaguar ambience. Wood and leather were retained wherever possible; otherwise it was all new.

The instrument display in front of the driver was a mix of digital and

THE FORTY ARRIVES

Period Jaguar advertising: "The Car You Always Promised Yourself—When I grow up!"

Although angular in styling, the XJ40 was always a very attractive car for its time with well-balanced shaping and size. This is a base 2.9-liter model from early in production.

Jaguar held a series of prelaunch events to hype the upcoming introduction.

It all started in March 1986, when home and overseas Dealer Council meetings were held to show off the new car. In April, there was a special viewing for all employees and then a major engineering presentation for the car at the Institute of Mechanical Engineers in London. In August, the trade launches took place at the Jaguar factory for UK and overseas dealerships and fleet buyers plus the UK Government.

This was followed by a series of "J-Days" held at the National Exhibition Centre in Birmingham, where all employees of Jaguar, their families and friends, plus dealers and families, component suppliers, and others—around 28,000 people in total—got the opportunity to see the new models.

Then there were the month-long international press launches, followed in October by a London City Financial viewing. On October 7th, initial cars were at the dealerships to coincide with the public launch of the car back at the NEC in Birmingham for the British Motor Show.

Another first for Jaguar was the creation of a unique television program devised to be viewed on UK TV on the evening of the public launch in October. *The Making of the Forty* was a major project filmed over a period of years covering the development and testing of the new car, in order to motivate the buying public.

analog. Traditional speedometer and rev counters were flanked on the left by vacuum fluorescent bar graph displays for fuel level, coolant temperature, oil pressure, and battery voltage, and on the right by a VCM (Vehicle Condition Monitor), a microprocessor driven unit presenting various diagnostic functions and visual and audible warnings. A total of 34 functions displayed through a dot-matrix screen. Before the instrument binnacle were controls for the onboard computer—providing information on speed and fuel consumption—and cruise control for certain models.

Most of the switchgear was new as were the steering wheel and stalks, and overall, regardless of model, equipment levels had improved. There was also a new sophisticated air conditioning system, optional on some models, offering not only temperature and air flow control but also humidity. A dashboard sensor allowed the system to react to external weather conditions.

Seats were all new, designed with the help of Loughborough University to maximize comfort. Cold-cure polyurethane foam was used in the seat cushions to provide consistent hardness and optimum support.

The all new center console incorporated the audio system and auxiliary controls. For the lesser equipped models, tweed upholstery was available instead of leather and straight grained woodwork instead of burr walnut.

The model line initially was quite small:

2.9-liter XJ6 (not sold in the States)

3.6-liter XJ6 (not sold in the States)

2.9-liter Sovereign (not sold in the States)

3.6-liter Sovereign (XJ6 in the States)

3.6-liter Daimler (Sovereign, later Vanden Plas in the States)

Service intervals had been dramatically reduced for the XJ40, even better than their German competition, and the starting price was a very competitive £16,495 for the 2.9-liter XJ6.

1987–1989: Inevitable Tweaks

With sales booming for the XJ40 in world markets, Jaguar continued to market the XJ12 Series models for the top-end buyers. Prices were increased by around 7 percent in 1987, justified supposedly by improved specification and high demand.

Prior to 1988, the audio systems in the XJ40 had been adapted from those fitted to other Jaguar models. From then forward, the top of the range models got a Clarion 926 HP unit.

Other trim and equipment changes involved heated door mirrors becoming standard on all cars; a rear sun blind and heated door locks became standard on Sovereign, Daimler, and Vanden Plas models; and 20-spoke alloy wheels became standard on Sovereign and VDPs.

In 1988, Jaguar sold 10,000 XJ40s in the UK alone (double the 1985 figure for Series 3 models). The model was voted The Boardroom Car of the Year.

Minor changes followed in 1989. Daimler and some early Sovereign models had displayed a matte-black rear panel to the boot, which was deleted in favor of normal body color paint. The door mirror design was slightly altered, and a new infrared remote door locking system was adopted.

Internally, Daimler and VDP models received boxwood inlays to the veneer, and Daimler rear compartment lighting was fitted to Sovereign models as well. Optional color piping was available for the interior leather trim. For the US and Canadian markets, the electrically operated metal sunroof (previously only standard on Daimlers/VDPs) was now fitted to the XJ6.

A new model, the XJR, was added to the range in 1989. Only fitted with the 3.6-liter engine, this car was clearly aimed at a younger market (see Chapter 9). Offered through a limited number of UK dealerships only, this sportier car boasted external body kits, Speedline 16-inch alloy wheels with Pirelli P600 tires, uprated suspension, and steering and interior trim changes.

At the British Motor Show in 1989, Jaguar debuted another new model, the 4.0 liter.

1990–1991: New Models

A major change took place for the 1990 model year with the introduction of the 4.0-liter AJ6 engine. The fundamental objective was to increase low-speed torque and enhance the feeling of effortless performance and refinement.

Capacity was raised to 3,980cc by increasing the stroke from 92 to 102 millimeters. In non-catalyst set-up, peak torque was increased from 249 pounds at 4,000 rpm to 250 pounds at 3,750 rpm. Revised camshaft profiles, tappets, valve timing, redesigned pistons, and a

The analog instrument pack fitted to all XJ40s from 1990 and carried forward to the later X-300 models.

The limited XJR model, which sought to create a younger and more sporting market for the XJ. Here in 3.6-liter form, it was later upgraded with the 4.0-liter engine and more changes.

crankshaft in forged steel instead of cast iron, all enhanced refinement.

A new more powerful engine-management system better regulated fuel and ignition and the new engine was set up to run on unleaded fuel.

The 4.0-liter engine used a new automatic transmission, the ZF 4HPE unit, featuring electronic controls linked to the engine-management system, and two programmable modes, "Normal" and "Sport." The manual Getrag gearbox was replaced by the 290 series with a repositioned gear lever. Internally, a three-plane gate pattern made reverse easier to engage. A larger diameter clutch and twin-mass flywheel came with the new gearbox.

With the 4.0 liter a new Teves anti-lock braking system was fitted, bringing the XJ40 in line with other Jaguar models.

To distinguish the new 4.0-liter XJ, more chrome trim was added to the rear area, with a wider range of paint finishes, and for the US market there were new headlights.

Resistance to the instrument binnacle prompted a new conventional analog instrument layout and ISO symbol warning lights instead of the Vehicle Condition Monitor. Some of the switchgear was also modified as was the sound-deadening insulation.

With the 4.0-liter engine came a revised XJR model from JaguarSport, with an extra 16 brake horsepower.

Model changes took place specifically for the US market as well. The XJ6 (Sovereign in the UK) got matte-black painted window frames, while the Vanden Plas got the same single headlamps as the UK cars. Specifically for the US market, cars would now be equipped with a unique front seat passive-restraint system that required significant engineering to install. Using a rail system above the front doors, the seat belts would automatically move and lock in position upon closing the doors.

Midterm in the production of the XJ40, for the US market, passive seat belts were added, at considerable extra development cost.

A limited-edition model was the Majestic, only available in a single Regency Red Mica paint, fitted with diamond turned alloy wheels with individual spokes faceted in red; the brakes were fitted with protective, color-coded shields. The car was trimmed in Autolux high-quality leather in magnolia, with contrasting Mulberry Red piping, red carpeting, and red overrugs. Mirror finished best burr walnut featured boxwood inlays and the car was equipped with everything as standard! This included a one-touch integrated security system that closed windows at the same time as locking the doors. The car was badged as a Vanden Plas Majestic with appropriate gold badging and cost $5,000 more than the best-equipped other model.

Also new for 1990 was an optional Sports handling pack, which provided stiffer springing, a larger diameter anti-roll bar, revised damper settings, lowered ride height, 16-inch wheels, a limited slip differential, and more direct steering, all of which improved the overall handling of the cars.

For 1991, there was another model change. Due to poor sales, the 2.9-liter car was deleted in favor of a new 3.2-liter model. Based on the 4.0-liter unit with the stroke reduced to 83 from 102 millimeters and the same bore, the capacity dropped to 3,293cc. Jaguar's first engine to be available only in a catalyst form, it produced a healthy performance increase to 200 brake horsepower, comparing favorably with the 3.6-liter engine.

External differences between the last of the 2.9s and the 3.2s came down to rear badging. Greater changes were made to all XJ40 models at this time. The original metric-sized wheels and tires were abandoned (apart from special order), replaced with conventional 7×15-inch wheels with 225/65 VR15 tires. The bottom-end models still retained steel wheels with plastic hubcaps, while the Daimler got a new style of wheel called Roulette. A fingertip lip to the fuel filler eliminated the need to push the filler to open it, and now the driver could operate all door locks from the driver's door.

By the end of 1991, over 158,000 XJ40s had been sold, marking the continued success of the XJ line.

1992: Mid-Life Crisis

More model changes followed, initially for the US market only. That special edition Majestic was offered in Black Cherry mica-metallic paint, sporting a chromed strip down the bonnet panel and chrome/rubbing strips along the sides of the car; they also received the Roulette alloy wheels from the UK Daimler model. Internally, the previous red support trim was now cream.

In mid-1992, at a time of growing competition from Lexus, Jaguar

improved their new car warranty period to three years on all models.

Engine upgrades during this period included twin cooling fans and revised camshafts and allied components to improve efficiency and decrease noise levels. The ZF gearboxes were recalibrated to improve shifts. A first gear inhibitor was incorporated allowing the engagement of first gear momentarily before engaging second gear; this helped traction in difficult conditions. A gearshift interlock required the driver to apply the brake before selecting park, and to have the car in park before the ignition key could be removed.

To help with increased electrical loads a more powerful battery was fitted, now situated in the boot. Catalytic converters were standardized on all cars.

Minor changes took place to the interior with better, more supportive seat foams, and different electric seat controls. Even the base XJ6 got power height-adjustable front seats. Door panels were redesigned with better storage, and there were revised switches.

1993: The V-12 is Dead, Long Live the V-12

Jaguar became the first British quality car manufacturer to fit air bags, with a redesigned steering wheel in the XJ40 in 1993. This was a standard fit for most markets including the US. Another safety feature involved revised front seat belts with web lock devices to prevent excessive tightening of the reels. At the rear, the seat belt buckle mountings were incorporated into the seating.

The air conditioning system was totally revised with more logical controls and a recirculating feature. Another new feature was an optional heated front screen.

New, fully integrated vehicle security and better hi-fi audio systems were fitted.

The XJ40 had been very successful for Jaguar, but by 1992 it still wasn't available in all markets around the world. Jaguar's new boss Nick Scheele had spent considerable time in the Far East in the hope of improving distribution in that area, and the company needed exposure in Eastern Europe and South America, plus China.

The culmination of Scheele's efforts meant that by 1993 XJ40s were selling in no fewer than 44 countries around the world with new importers in the Czech and Slovak Republics, Hungary, and Slovenia. China also got to see the XJ40 for the first time in 1993.

By 1994, Jaguar had seen a 43 percent increase in sales over the previous year in the US alone. Further gains were a 100 percent increase in Singapore, 40 percent in New Zealand, 55 percent in Sweden, and the first 40 cars had been delivered to China.

(OPPOSITE) Finally, in 1993, the XJ40 body was redesigned to take the V-12 engine. Here, seen in Jaguar XJ12 form, it used the four-headlight treatment normally applied to lower specification models. Unique features at the time included the black radiator grille vanes and gold badge.

Midterm US range of XJ40s from the Sovereign in the foreground to the Majestic at the rear. Note the different alloy wheel and window surround treatment, according to model.

Despite all the changes up to this time, the old Series 3 V-12 cars had still been selling. Bringing production to an end left Jaguar with no direct replacement. Because the XJ40 had been designed *not* to accept Rover's V-8, wedging in a V-12 meant significant modifications. Over £35 million was spent redesigning the XJ40 to accept Jaguar's V-12, but part of that figure was offset by the need, afterward, for only one production line.

Accepting the V-12 meant 44 percent of the total XJ40 paneling had to be changed or modified. Sixty changes were attributable to the front end of the car alone. To support the larger engine, a new subframe of a different open form was designed—which in future years would prove to be longer lasting than that used on the six-cylinder models. Springs and dampers were uprated to handle the extra weight.

The V-12 engine, now at 6.0 liters (see Chapter 9), was mated to a GM 4L80E automatic transmission and never received a manual gearbox. The 6.0-liter XJ40 outperformed the departing Series 3 V-12s.

Two variants were offered, the Jaguar XJ12 (fitted out to Sovereign specification except for four headlights) and the Daimler Double Six/Vanden Plas.

1994: More New Models, But the End Is Near

Announced for the 1994 model year, the 3.2-liter S (for Sports) offered sportier handling at a competitive price with a more youthful exterior and interior finish.

External changes were subtle with color-coordinating grille vanes, headlight surrounds, front valance and door mirrors. At the side a "3.2 S" badge adorned the lower front wings and along with the conventional swage coachline, there were two more coachlines at a lower level. New 7×16-inch five-spoke alloy wheels finished in eggshell or silver were fitted with 225/60 ZR 16 tires. At the rear

(**ABOVE**) The XJ6 Gold, a limited edition last-of-the-line model to create an attractive luxury entry model to the range.

(**LEFT**) The sporty 3.2-liter S (later also in 4.0-liter form) was produced to entice a younger audience to Jaguar.

red or gray neutral density lighting was used along with a plastic in-fill panel to the boot lid. The 3.2 S was offered in a limited range of colors.

Internally, walnut veneer was replaced by rosewood stained maple with dashboard top roll and door tops finished to match the seat trim. The seat facings were unique with horizontal pleating and mulberry red stitching.

Mechanically the car was standard except for the Sports pack. Priced for the UK market only, the 3.2 S cost just £2,000 more than a standard 3.2-liter XJ6.

A few months later, the 4.0 S was announced, and apart from the larger engine and new badging, was exactly the same as the 3.2 S.

Release of these two models was paired with deletion of the base model 4.0-liter XJ6, as was the 3.2-liter Sover-

eign. The remaining models received a broad range of changes from new wheels to myriad trim updates.

All XJ40s from 1994 were fitted with a passenger side airbag, which meant the deletion of the glove box in all cars.

With improved production methods and quality control, Jaguar was able to offer extended 10,000-mile service intervals.

It wasn't over yet for the XJ40, as yet another new model came into being in 1994, the Gold, to boost the entry level range, which now offered three models in the under-£30,000 price bracket.

Aimed at a limited-edition market, the Gold represented excellent value for money at £28,950. It was based on the 3.2-liter engine car with leather trim, burr walnut veneer, uniquely fluted

seating, and color matched gearshift and handbrake surround.

Outside, 16-inch Kiwi alloy wheels, gold on black wheel centers, gold radiator grille badge, twin gold coachlines, and rear badging completed the look. Offered in a limited range of colors, the model was a challenge to the Mercedes E280 and BMW 730i.

The XJ40 did well for Jaguar and its Ford owners. A total of 208,706 XJ40s had been produced and sold in just 8 years.

1994–1995: A New Series

Over £50 million was spent modernizing the assembly facility at the Browns Lane Plant in Coventry in readiness for a new model, the first new car produced under Ford ownership. The XJ saloon

VARIATIONS ON THE THEME

The special order long-wheelbase Majestic model, not to be confused with the up-market Majestic special edition produced for the States.

In addition to the US Majestic model, the name was used for another XJ40-based car incorporating a 125-millimeter increase in the wheelbase to provide extra room for rear-seat passengers. To achieve this, the floorpan was extended, new rear doors were made, and the roof was continued back at a slightly higher level.

All the fabrication work was carried out from standard production bodies by Project Aerospace in Coventry, after which Jaguar finished and painted the body and did final trimming in Jaguar's Special Vehicle Operations department.

Many of the Majestics built to special order were fitted out to Insignia trim (see Chapter 9).

A number of XJ40s were custom-equipped by Jaguar's Special Vehicle Operations Department with special paint finishes and enhanced interior trim, like this Daimler model.

had been Jaguar's biggest money earner, so the vehicle code named X-300 was introduced at the 1994 UK motor show as the "New Series XJ."

Stylistically the aim was to bring back curvature, capitalizing on the heritage of the earlier XJs, while retaining and improving the quality and enhancement achieved with the XJ40. Underpinnings of the body derived from the XJ40, but significant new paneling was in store for the X-300.

At the front, the classic four headlight treatment served all models, now with new surface reflectors giving 15 percent more light output. Auxiliary lights were now standard on all models and the indicators were incorporated in the polyurethane injection molded bumper styled with chrome finishers. A traditional Jaguar radiator grille was neatly blended into the more rounded front of the car, finished differently dependent on model. The hood was sculpted around the headlights with a pronounced hump in the middle.

From the side view, it was very much XJ40 although substantially repaneled with, for example, the rear wing being a single panel including the E pillar to eliminate the need for a finisher. Window surrounds and other detail trim were also altered according to model, but all bodies featured not only coachlines but below-level rubbing strips finished to body color.

(**ABOVE, LEFT**) The X-300 XJ6 and XJ12 range covering base models with a 3.2-liter engine, through to the top-of-the-range 6.0-liter, brought a return to a more curvaceous style.

(**LEFT**) The new curvaceous styling of the X-300 was well received and took Jaguar back to a period of styling success in the 1960s and 1970s.

Jaguar's first production supercharged car, the XJR, was one of the fastest saloons in the world, aimed at the BMW M market. This was the first model to feature the mesh radiator grille, which has been used ever since for many Jaguar models.

The newly appointed X-300 interior was adapted in various forms to suit the different models and their probable owners. This is a Sport with a large degree of black finish and, in this case, half leather trim.

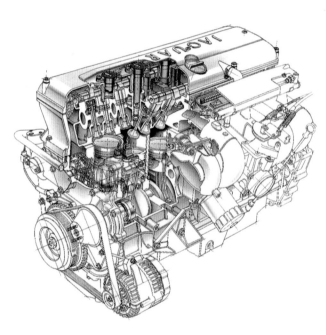

The new AJ16 multivalve, coil-on-plug engine in 3.2-liter and 4.0-liter form for the X-300 models, also available in a supercharged version.

The new rear wings and boot panel followed the "curvature" approach with new lighting, registration panel treatment, and a polyurethane bumper as at the front. Unusually, for the US market, a revised boot pressing was used to accommodate American style registration plates.

Overall the X-300 required 11 percent fewer body panels and sub-assemblies. Fit and finish were drastically improved on all panels, and even the door gaps were 25 percent reduced compared to the XJ40. The rear doors incorporated a "jiggle" on the leading edge, so the front doors overlapped the rear doors on closure. All glass areas were semiflush glazed, so the torsional rigidity of the glazed body was improved by 18 percent. All exposed body panels were double-sided zinc coated to offset corrosion.

With the X-300 Jaguar was the first manufacturer to offer total electrical short circuit protection with no fewer than 90 fuses, gold plated connec-tors, shielded leads, and more robust installation.

Internally, most was new. There were new seats and frames, door panels, carpeting, woodwork, and other trim, all of improved quality. While the main instrument pack was unchanged, accompanying switchgear and trim was altered.

Mechanically the new cars were also vastly improved. Jaguar installed the AJ16 six-cylinder engines, also used in the later XJS models. Instantly recognizable by the silver finished cam cover and concealed wiring, the new engine was available in both 3.2-liter and 4.0-liter capacities, plus for the first time a supercharged 4.0-liter version for a new model, the XJR.

The AJ6 units were improved with over 100 newly tooled engine and ancillary components. There was a new cylinder head and reworked block, camshaft profiles, on-plug coils, a higher compression ratio, new pistons, sequential fuel injection timing, knock sensing, and new valve gear. Also debuted were a new throttle system, induction manifold and exhaust, plus a highly sophisticated engine-management system designed to comply with the latest American onboard diagnostics legislation. All these changes brought between an 8 and 10 percent increase in performance over the AJ6 units.

The supercharged 4.0-liter engine developed 326 brake horsepower and 378 pounds-feet of torque at 5,000 rpm. A Roots-type Eaton M90 mechanically driven supercharger provided immediate throttle response. A seven-rib poly-vee belt drove the supercharger off the engine at 2.5 times crankshaft speed for low speed torque. Valve timing was changed to suit the lower compression ratio. The inlet system was unique to the supercharged engine with a heat exchanger of the intercooler system incorporated into the manifold.

The 6.0-liter V-12 was still available on selected models, now taking advantage of the same new engine management system.

To improve off-line performance and decrease weight, the existing four-speed automatic transmission was reengineered with a reduced-size torque converter. The five-speed manual remained an option for six-cylinder models.

A new brake actuating system provided improved pedal effort. Rear brakes were now ventilated, with larger calipers. There was also a new traction control system and ABS was standard across all models. A new electronically controlled ZF rack-and-pinion steering system was employed.

The suspension was basically carried forward from the XJ40 models, but in line with the policy for the "New Series," each model would be differentiated by not only trim, but also ride. With no fewer than five suspension packages according to model, each was designed to match the characteristics for each car.

X-300 RANGE STRATEGY

- XJ6: Entry-level models aimed at cost-conscious traditional individual and fleet buyers, and new conquest sales from other brands
- Sport: Aimed at younger, sportier owners
- Sovereign: Those buyers looking for luxury appointments and quality finishes
- XJR: For the ultimate performance and driving experience
- Daimlers/Vanden Plas: Aimed at the company chairman/president, offering the ultimate in luxury and comfort

RIDE PACKAGES

XJ6 3.2-Liter, 3.2-Liter Sovereign, 4.0-Liter Sovereign, and Daimler Six

- 20 percent stiffer front anti-roll bar
- Touring damper settings
- 7×16-inch wheels with 225/60 ZR16 tires

XJ Sport 3.2- and 4.0-Liter

- Unique front and rear anti-roll bars
- Unique damper settings
- Reduced right height by 10mm
- 8×16-inch wheels with 225/55 ZR16 tires

XJR

- Unique uprated front and rear anti-roll bars (85 percent)
- Unique front and rear springs
- Unique damper settings
- Further reduced ride height
- 8×17-inch wheels with 225/45 ZR17 tires
- Limited-slip differential.

XJ12

- Unique front and rear anti-roll bars
- Unique damper settings
- Softer springing
- Reduced ride height
- 8×16-inch wheels with 225/55 ZR16 tires

Daimler Double Six/ Vanden Plas

- Touring springs and damper settings
- 7×16-inch wheels with 225/60 ZR16 tires

1996–1997: The Long and the Short of It

For 1996, Jaguar reengineered the front passenger area to incorporate a new glove box below the air bag area and even produced a kit for existing owners to convert their early X-300s.

They also introduced a 4.0-liter version of the base model XJ6, responding to customer demand. More significant were longer-wheelbase bodies made available for Sovereign, Daimler, and Vanden Plas models, providing an extra 6 inches of space in the rear compartment.

In 1996, Jaguar recognized the centenary of Daimler, which it purchased in 1960, with two special Daimler Century models. Limited to 100 Daimler Six Century (six cylinder) and 100 Daimler Double Six Century (V-12) models, the cars were equipped with every conceivable extra and on the long-wheelbase body, becoming the most expensive production cars the company had ever produced.

Yet another new model appeared later in the year, the Executive, aimed at a younger businessman. Based on the standard 3.2-liter XJ6, it featured sports leather trim, half wood/leather steering wheel, the wider Sport wheels and tires, plus air conditioning.

Throughout the later production period, the longer wheelbase could be specified for other X-300 models.

In April 1997, Jaguar ended V-12 production after just 3,492 X-300 V-12s were produced.

The rest of the X-300 range only stayed in production until late 1997 as a stop-gap measure until Jaguar released its new AJ-V-8 engine, for which this car was destined. Despite the short production period, a total of over 92,000 X-300 XJs were produced.

The range topper limited-edition Daimler Century, one of just 200 produced to commemorate the centenary of the Daimler Motor Company. All examples used the long-wheelbase bodyshell and were finished with every conceivable extra and some unique badging and trim.

The sumptuous interior of a Daimler Century model.

1997–2002: Into V Formation

Late in 1997, Jaguar announced their new saloon car range, coded X-308, and known as XJ8. These incorporated the new AJ-V-8 engines, first seen a year before in the XK8 sports (see Chapter 11), now available in 3.2 liter, 4.0 liter, and 4.0-liter supercharged form. All engines were matched to a ZF five-speed electronically controlled automatic transmission and, due to a general falloff in demand over the

The V-12's run ended in 1997. Here, the very last X-300 XJ12 is set against other Jaguar models that used a V-12 engine. From left to right: the prototype sports/racing XJ13, E-Type Series 3 roadster, one of the very first XJ12s (Series 1), European Touring Car Championship winning XJ-S, and Le Mans–winning XJR.

years, no manual gearbox versions would be available.

Most of the chassis development technology came from the XK8 sports. The new saloons featured a front suspension system of spring to beam, twin wishbone design, and for braking, the latest version of the Teves system. Stability and traction control, variable ratio speed-proportional power steering, drive-by-wire throttle management, and retuned rear suspension all made for better performance, handling, and comfort. Allied to this a new on-centerline differ-

ential and two-piece propshaft helped reduce noise and vibration.

Electronically the X-308 models incorporated multiplex harnessing to inter-communicate and share information and a controller area network for all drivetrain functions.

Environmental issues were also addressed. Waterborne paint systems were used, cutting down solvent emissions by 85 percent and over 80 percent of the car was now recyclable.

Body changes from the X-300 were subtle with more rounded bumpers and

new lighting, while the interior came in for a major revamp, including an entirely new dashboard with the wood veneer attached to light aluminum substrates, new instrumentation and switchgear, much of it from the XK8 sports, a new center console, steering wheel with many controls mounted around it, and better quality single-piece door cards.

The passenger-side glove box had been improved in size, the seat travel extended by 20 millimeters for better legroom, and there was another improved audio system.

(ABOVE) Comparisons between the X-300 and X-308 models. Subtle changes included a new bumper bar with split chrome blades, new lighting, and a less angular radiator grille.

(RIGHT) One of the AJ-V-8 engines, a 3.2-liter version as fitted to many XJ8s at this time.

(**ABOVE**) An entirely new interior for the X-308 XJs with wood veneer set onto an aluminum substrate for lightness and better fit, new instrumentation (taken from the XK8 sports car), new seating, and other trim.

(**LEFT**) Interiors for the XJ8 varied according to model, with different wood treatments, degree of black trim, and even seat design.

The XJ8 3.2-liter Sport model with Dimple alloy wheels, marginally reduced ride height, and a sportier interior.

The model range upon introduction was XJ8 3.2, XJ8 4.0, XJ8 3.2 Sport, XJ8 4.0 Sovereign, XJR, Daimler Eight, and Daimler Super V-8 (Vanden Plas in the States). Short- and long-wheelbase policy followed the same practice as the earlier model.

There were various minor changes to the specification and model range throughout the relatively short pro-duction period of the X-308, including replacement of the base model with the Executive, and the Sovereign becoming the SE (Special Equipment).

A total of 126,000 X-308s were produced and although it outsold the X-300 for only one year, overall it sold steadily and was a well-respected and refined Jaguar saloon. But it was time for something entirely new.

The special Jaguar XJR 100, all of which were produced in black and commemorated the centenary of Sir William Lyons, the founder of the Jaguar (previously Swallow).

SPECIFICATIONS

MODEL	2.9-liter XJ40	3.2-liter XJ40	3.6-liter XJ40	4.0-liter XJ40	6.0-liter XJ40
ENGINE SIZE	2,919cc	3,239cc	3,590cc	3,980cc	5,993cc
CARBURETION	Fuel Injection	Fuel Injection	Fuel Injection	Fuel Injection	Fuel Injection
MAXIMUM BHP	165@5,600	200@5,250	221@5,000	235@4,750	318@5,400
MAXIMUM TORQUE	176@4,000	220@4,000	249@4,000	285@3,750	342@3,750
GEARBOX	5-speed	5-speed	5-speed	5-speed	n/a
AUTOMATIC	4-speed	4-speed	4-speed	4-speed	4-speed
0 TO 60 MPH	9.6 sec.	8.5 sec.	8.8 sec.	7.6 sec.	6.9 sec.
STANDING ¼ MILE	n/a	n/a	18.9 sec.	n/a	n/a
TOP SPEED	118 mph	118 mph	135 mph	138 mph	155 mph
AVERAGE FUEL CONSUMPTION	19.8 mpg	19.8 mpg	18.7 mpg	19 mpg	15 mpg

MODEL	3.2-liter X-300	4.0-liter X-300	4.0-liter S/C X-300	6.0-liter X-300
ENGINE SIZE	3,239cc	3,980cc	3,980cc	5,993cc
CARBURETION	Fuel Injection	Fuel Injection	Fuel Injection	Fuel Injection
MAXIMUM BHP	219@5,100	249@4,800	326@5,000	318@5,350
MAXIMUM TORQUE	232@4,500	289@4,000	378@3,050	353@2,850
GEARBOX	5-speed	5-speed	5-speed	n/a
AUTOMATIC	4-speed	4-speed	4-speed	4-speed
0 TO 60 MPH	8.9 sec.	7.8 sec.	6.6 sec.	6.8 sec.
STANDING ¼ MILE	n/a	n/a	14.9 sec.	n/a
TOP SPEED	139 mph	144 mph	155 mph	155 mph
AVERAGE FUEL CONSUMPTION	26.9 mpg	26.8 mpg	23.4 mpg	18.4 mpg

SPECIFICATIONS

MODEL	3.2-liter X-308	4.0-liter X-308	4.0-liter S/C X-308
ENGINE SIZE	3,248cc	3,996cc	3,996cc
CARBURETION	Fuel Injection	Fuel Injection	Fuel Injection
MAXIMUM BHP	240@6,350	290@6,100	370@6,150
MAXIMUM TORQUE	233@4,350	290@4,250	387@3,600
GEARBOX	n/a	n/a	n/a
AUTOMATIC	5-speed	5-speed	5-speed
0 TO 60 MPH	8.1 sec.	6.9 sec.	5.3 sec.
STANDING ¼ MILE	n/a	14.8 sec.	13.5 sec.
TOP SPEED	140 mph	150 mph	155 mph
AVERAGE FUEL CONSUMPTION	23.5 mpg	23.7 mpg	21.6 mpg

CHAPTER TEN

THE RETURN OF THE XK

X-100 to X-150

Jaguar's postwar XK120/140/150 and E-Type perfectly met the sports car needs of their era. The later XJ-S met the requirements for something more sophisticated— a grand tourer with sports car agility. But even with its successful revamps, it had been in production for too long. With the new X-300 saloon well under development, Ford committed to develop a new sporting model. This entirely new car came to be in just five years.

1991–1997: The Cat Is Back

Work commenced on the X-100 back in 1991 and by the following year styling concepts were produced in-house and from Ford and Ghia. From these the final exterior and interior designs led to a fiberglass mock-up that was shown to the US dealership network in 1993, receiving excellent feedback.

Although Ford was ready to commit financing to the project, much of it would go to develop the new AJ-V-8 engine (see below). Any new sports car would thus have to be based on the existing XJ-S floorpan. Now the hurdle for Jaguar engineers was to develop a new

suspension around it. Using the existing XJ-S components would have been cheap, but technology had moved on.

For the front, a conventional double unequal-length wishbone design combined with coil springs and telescopic shock absorbers was chosen for the XK8 with the road springs mounted directly to the body and spring rates developed to create less stress. Hydraulic mounts, filled with oil, were set between the front crossmember and engine, tuned to ensure spring loads remained isolated.

Because it was no longer possible to utilize the existing XJ-S subframe, Jaguar

designed a new lightweight aluminum subframe built to aircraft-quality standards. It weighed a mere 15.8 kilograms and was treated with Dacromet to guard against corrosive interaction with steel components.

At the rear, an adaptation of the then new X-300 suspension with an A-frame and monostrut design using a pendulum arrangement was chosen.

An extra-cost option (initially only for the coupe) was Jaguar's CATS (Computer aided Technology System). This optimized ride and handling under electronic control, sensors

With the XK8, launched in 1996 in both convertible and coupe form, Jaguar returned to more curvaceous styling. An entirely new look was built around the existing XJS floorpan.

front and rear monitoring all movements.

A vastly improved braking system with larger front and rear (28×305-millimeter) ventilated discs was used, along with the established Teves anti-locking system, and Automatic Stability Control, also acting with Traction Control.

For steering, Jaguar adopted the ZF Servotronic speed-sensitive rack-and-pinion system with a variable ratio feature to benefit low-speed maneuverability and high-speed stability.

Jaguar spent £160 million developing the new engine for all models (which included setting up the production facilities at Ford's Bridgend factory in Wales). The parameters were excellent performance with refinement, a high standard of quality and durability, reasonable cost of ownership, as well as compliance with current and anticipated future legislation over emissions, fuel efficiency, and noise levels.

The engine design came directly from Jaguar engineers. The basic configuration decided upon was a V formation for compactness and eight cylinders

for refinement. The engine was of a four-camshaft design with a 90-degree V formation. Of square cylinder dimensions (86 millimeters), and with a capacity of 3,996cc, it was rated at 290 brake horsepower at 6,100 rpm. Jaguar claimed the highest specific output and torque per liter of any production engine.

The engine block, designed for lightness, was stiff and torsionally strong with cast-ribbed web-connected banks and a closed-deck design. To cut back on weight, there were no conventional iron cylinder liners. The bores were treated with Nikasil (a nickel silicon carbide),

The highly refined Jaguar AJ-V-8 found its first home in the XK8 and became the mainstay of Jaguar engine production until 2009.

developed in Formula One. Cosworth produced the unique twin cylinder heads, heavily ribbed for stiffness. There was a five-bearing crankshaft produced from spherical graphite cast iron, with connecting rods forged by highly accurate Krebsoge powder-sintering technique. Alloy flat top, short skirt pistons were fitted in conjunction with a high 10.75:1 compression ratio.

The new engine incorporated variable cam phasing on the inlet cams with a 30-degree range of adjustment. A four-valves-per-cylinder configuration was adopted with slender diameter shafts operated by low-mass aluminum bucket tappets. The chilled cast iron camshafts were drilled from end to end to save more weight. Camshaft drive was by four single-row chains to minimize the depth of the engine and provide greater safety in service. A 28-degree angle between the inlet and exhaust valves facilitated a narrow squish-free pent-roof combustion chamber.

Close-coupled catalytic converters were fitted along with thin walled manifolds improving emissions, and to ensure a very quick warm-up of the engine a patented low-volume split-block cooling system was devised (for which Jaguar received an award). The V-8 engine could achieve a warm up in less than 4 minutes from cold.

Additional weight reduction was achieved by the extensive use of plastics including the intake manifolds with an integrated fuel rail to reduce complexity and improve injector targeting.

A Nippondenso 32-bit electronic engine-management system was the best produced at the time in terms of optimizing throttle adjustment with improved idle speed control and better emissions.

No manual gearbox was offered on the XK8, and ZF supplied their then new five-speed 5HP 24 automatic transmission. It operated through Jaguar's new "J-gate" selector and the unit was supposedly a sealed-for-life gearbox.

Although standard fit were 7J×17-inch alloy wheels with Pirelli P-Zero tires with an asymmetric tread pattern, there was also the option of 8J×18-inch wheels and tires with, for the first time on a Jaguar, larger-section wheels/tires at the rear.

The XK8 also accommodated the very latest state-of-the-art security system.

Although utilizing the existing XJ-S floorpan, the body was all new with no carryover of any external paneling. With a prominent "nose" instead of a grille, and aerodynamic front and integrated high-tech lighting, the proportions were remarkably similar to Jaguar's 1960s E-Type. The one-piece forward-hinged bonnet even had a center power-bulge like the E-Type. The beautifully sculpted bodywork around the sides was unadorned with bridgework. Wide opening doors with flush handles had frameless windows that closed tight automatically against the high-tech rubber seals when the doors were shut.

At the rear, the curves continued with plastic bumper/under valance,

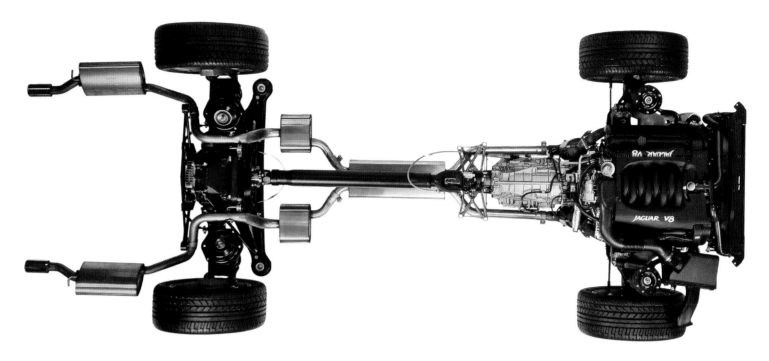

A new engine, transmission, and suspension, plus aircraft-technology subframe for the XK8.

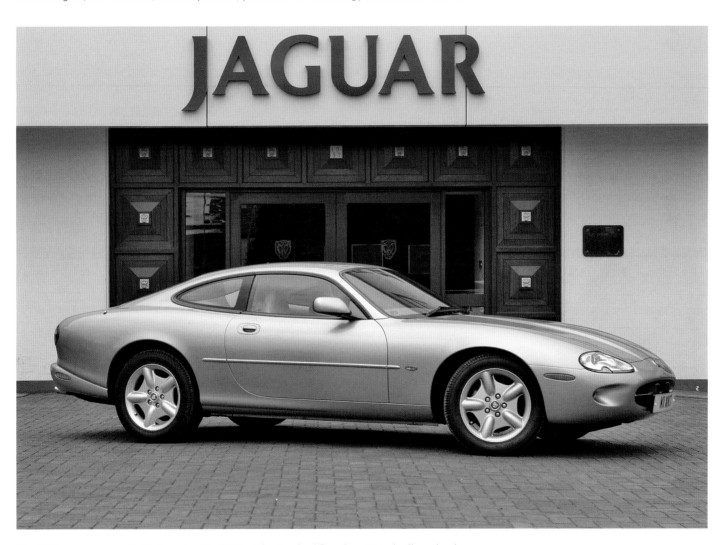

The XK8 coupe as originally launched in 1996 with standard Revolver 17-inch alloy wheels.

Characteristic Jaguar interior with the traditional touches of wood and leather mixed with modern technology. Notice the sculpted seats, whose covers hide the previous model (XJS) underframes.

wraparound lighting, and a conventional boot lid, even on the coupe, giving access to a large boot area. The boot had been designed to accommodate two sets of golf clubs (thought a necessary attribute for many markets!).

The XK8 was launched as both a 2+2 coupe and a 2+2 convertible, the latter with an electrically controlled, lined top but one that would retract down to an area behind the rear seats. It did not have an electrically controlled hard cover. Instead, to maintain luggage space in the boot, Jaguar supplied a matched-leather tonneau cover to be manually fitted when the top was down.

Internally, the XK8 was as luxuriously equipped as any saloon. New electronic instrumentation appeared with the speedometer and rev counter still in front of the driver but now deeply inset

into the dashboard with auxiliary instruments in their usual position, centrally mounted. The center console was similar in design to that used in the saloons and all models featured both driver and passenger air bags.

The seating was of a new design, but cleverly devised around the old XJ-S seat frames. The rear seating arrangement was rather cramped for adults, but necessary for some markets.

Two interior trim options were offered, the Classic and Sport. The former featured all leather plus walnut veneer; the Sport had part leather/part cloth upholstery and dark stained maple woodwork.

After an initial public preview at the Geneva Motor Show in March of 1996, the car made its full debut at the London Motor Show in October, where

it was met with immediate acclaim. In 1997's first quarter, Jaguar achieved their best-ever sports cars sales with 3,977 cars delivered (a 257 percent increase over 1996). By the end of the year, sales were 14,619—50 percent of which were in the States.

1998: R-Rated

It was inevitable that a more powerful version of the XK8 would come, as the XJR (supercharged saloon) had been so successful. The XKR went on sale in May 1998 in both coupe and convertible forms, the most powerful Jaguar offered up to that time. Initially the XKR was available in only a limited range of colors including a new (unique to the model) Phoenix Red. The limited colors arose from fears that certain paint pigments

(ABOVE) The XKR in its launch color of Phoenix Red, set against the backdrop of previous Jaguar sports cars, from left to right, XKSS, XK150, XK140, and XK120.

(LEFT) The 4.2-liter version of the V-8 engine, here seen in supercharged form, mated to the new six-speed transmission.

THE XK THAT NEVER WAS

The concept XK180, while not making it into production, yielded several features to be seen later on XK models.

At the Paris Motor Show in 1998, Jaguar unveiled their XK180, significantly based on the XKR. Produced by Jaguar's Special Vehicle Operations Department, the body was entirely of aluminum with a shortened wheelbase and following styling inspiration from Jaguar's D-Type sports/racing car of the 1950s. Much of the mechanicals came from the XKR but with racing-style Bilstein aluminum shock absorbers mounted inside the springs. An uprated Brembo racing brake system was adopted with larger cross-drilled discs and aluminum four-pot calipers. The car used higher geared rack-and-pinion steering and all bushings and frame supports were strengthened.

The Mercedes transmission was retained but with a "slick shift" operation. Supercharger gearing was increased by 10 percent with a bigger intercooler, improved induction, and big-bore exhaust system. The end result was a reported 450 brake horsepower with 445 pounds of torque. The car was fitted with 20-inch split-rim alloy wheels.

The whole project took a mere 12 months to complete and had Jaguar had the time and resources to make the project meet road legislation, it might have reached production. Just two examples were produced, both still in existence; some elements would later be found in production XKs.

would fade under the heat generated by the supercharged engine.

At a casual glance, the XKR looked no different from the XK8 but there were subtle changes. At the front, a stainless-steel mesh grille was fitted inside the nose and a "Supercharged" badge appeared in the center of the nose top panel (the same appearing on the boot lid). The hood carried two louvered areas to aid airflow and help cool the engine bay. At the rear, a discreet spoiler was fitted to the edge of the boot lid, also to aid airflow, and there was "XKR" badging. To handle the increased performance, 18-inch alloy wheels were standard.

Internally, little changed, with the Sport pack being standard. XKR owners got standard CD auto-changer, cruise control, all-leather faced seats, memory mirrors, and a headlamp power washer. XKR logos were embossed in the center of the half wood/leather steering wheel along with a "Supercharged" legend in the main instrumentation cluster. The rev counter was recalibrated to match the new engine.

The engine was a direct implant from the XJR saloon released a year earlier, with only minor changes to accommodate the engine. The ECUs were recalibrated to a sports car spec, as the engine provided an extra 28 percent in power over the normally aspirated XK8. An even more efficient radiator and condenser were fitted along with a modified exhaust system, with lots of other minor changes in pipework and fitting, including heat shields, to meet the needs of the more powerful engine.

To handle the extra power a Mercedes WSA 580 five-speed electronically controlled and intelligent automatic transmission was used. Also, the XKR was the first car in the world to be fitted with the ZF Servotronic Mark 2 power-steering system, mated to a two-piece shaft.

Suspension was beefed up with uprated front discs and CATS.

1999: Changes

With sales still very buoyant, various minor changes took place. That XKR Phoenix Red color became available, along with others, for all models. To improve the convertible's structural integrity, a stiffening brace was fitted between the seat belt pillar and the B-post. Also, an X-brace was fitted to the front to aid removal and fitting of the radiator and air conditioning condenser. Soon other strengthening had to be made to adhere to new European legislation. Reinforced panels extended the driver and passenger toe-boards along with internal stiffening to the adjacent side members.

By October, the Servotronic 2 steering system was adopted for the XK8 as well as the XKR, and all the V-8 engines got air assisted fuel injectors, improving performance.

At the Frankfurt Motor Show in 1999, R-Performance options became available, derived from the work done on the XK180 concept. The BBS company supplied new split-rim alloy wheels in 19-inch and 20-inch sizes. With 20-inch wheels, the XK needed rubber extensions to rear wheel arches to meet construction and use regulations. Brembo supplied new two-piece cross-drilled, ventilated discs of 28×330 millimeters with aluminum four-pot calipers, and there was also a handling pack that could be fitted to those cars with CATS suspension.

2000: Better Spec and the First Special Edition

Adaptive Cruise Control was now available, standard on XKRs, and an integrated DVD-based satellite navigation system could also be specified from 2000, along with an upgraded ABS system and rain-sensing wipers. Better audio systems were also available.

The biggest mechanical change, albeit not visible, was the adoption of conventional steel cylinder liners, as the Nikasil coating failed under certain conditions.

In April, the Silverstone special edition was announced, finished in Platinum Silver. Twenty-inch Detroit alloy wheels were fitted, badging was amended, and the interior was finished in black with red stitching with matching carpets and gray stained maple woodwork. The model also got the top-of-the-range Alpine audio system. Silverstone coupes also got the handling package mentioned previously.

2001: Improved Specs and Another Special Edition

External trim changes included flush-mounted auxiliary lights at the front and a new bumper to accommodate them. A wider choice of alloy wheels was now available and jeweled rear lighting was designed for the car with chromed surrounds. There was also a new rear bumper to hide more of the car's underside. The boot lid had a push-button release (previously a key had to be used) and there was a chromed plinth on the boot lid.

Side airbags were fitted to all models, along with the latest Adaptive Restraint technology. Internally the seats were completely redesigned with separate headrests, more supportive bolsters, and 12-way electric adjustment. A new dual-band in-car telephone system was also provided.

Another special edition, the XJR 100, was produced to commemorate the centenary of the company's founder, Sir William Lyons. All 100 cars were finished in black with black interior, instruments got chromed surrounds, and the cars featured R-Performance handling kits.

2002–2003: The "New Generation" XK

This was a year of significant change to improve failing sales for the XK. The V-8 engine was increased to 4.2 liters, improving overall performance and economy. On the XKR, the supercharger was fitted with helical rotor gears cutting down noise and spinning 5 percent faster. Now the XK8 produced 300 brake horsepower, and 400 brake horsepower for the XKR.

A new six-speed ZF 6HP 26 automatic transmission specifically designed for rear-wheel-drive sports cars was now standardized on all models—smaller and lighter with greater torque capacity. A revised rear axle ratio was adopted to match the new gearbox.

Brembo brakes became standard for the XKR, enlarged brakes for the XK8. All models now incorporated an Emergency Brake Assist (EBA) providing extra braking pressure under severe conditions. Dynamic Stability Control (DSC) was standard.

More exterior changes took place with Xenon headlights standard on the XKR along with automatic headlight leveling and an auto-illuminate in poor ambient light.

Inside, dual color-trim options became available as did factory-fitted Recaro seating and alloy pedals and other trim. Finally, the side rubbing strips were eliminated.

Yet another Special Edition, the Portfolio, was released only for the US market and only available as an XKR convertible. Unique exterior paint finishes included Coronado or Jupiter Red, with 20-inch Detroit alloys. The interior was similarly finished with Recaro seating and heavily figured veneer two-tone upholstery.

(ABOVE) The facelift XK8, instantly
identifiable from the sill extensions,
broader front bumper, and wider mouth.
Note the side rubbing strips are absent for
a cleaner look.

(RIGHT) The later, better-equipped
interior of the XK8. Earlier dial
instruments on the center dashboard have
been replaced by an integrated satellite
navigational system.

The XK8 S for the European market and the Victory for the States were well-equipped special models.

2004: Further Changes and Another Special Edition

The most significant changes to the XK came about at this time to give the car a more contemporary feel.

At the front, there was a completely new nose section with a deeper mouth and lowered front giving the car a more aggressive look and improved aerodynamics. This incorporated a new splitter bar (and mesh grille for the XKRs), "overriders," and registration plate mounting.

Plastic sill covers lowered the body level to match the front; at the rear was a new bumper. These touches and a redesigned spoiler gave the car a slightly squatter look. Larger exhaust tailgate for the XK8 and a larger spoiler, and quad tailgates for the XKR, with more wheel options, completed the exterior package.

The main structure and exterior paneling remained unchanged but there were revised paint colors, and black window surrounds (instead of gray) featured on all models.

The interior was refreshed with the option of two new wood veneers, Elm and a high-polished Piano Black. The R-Performance alloy pack for instrumentation was now also available for all cars.

A new Automatic Speed Limiter (ASL) prevented the car from exceeding a preset speed limit.

In September 2004, Jaguar announced that they had sold a record 40,000 XKs in the States, making it the company's most successful sports car ever.

Yet another Special Edition was announced in 2004, the Carbon Fibre, exclusively for the UK market. Based on the XKR, it was the best equipped XK yet. With touches like a carbon-fiber finished dashboard, the Special Edition cost around £4,500 more than any other XK at the time.

2005: Swansong

At the Geneva Motor Show that year, Jaguar announced their final versions of the XK, the 4.2 S. A similar vehicle was launched at the Los Angeles Auto Show, called the Victory. Available in both XK8 and XKR (coupe and convertible) they had a short production run of 1,050 cars. Mechanically standard, externally they were identifiable by badging on the boot and the growler/checkered-flag emblem, 19-inch Atlas alloy wheels, and a new color palette with darkened rear light clusters.

Internally, sill treadplates were highly polished with the checkered-flag emblem, wood areas were finished in elm (XK8) and carbon fiber (XKR), there was alloy finishing to instruments, and the carpets included appropriate model emblems. Soft grain leather was chosen, along with all the usual extra-cost options including Bluetooth connectivity and a Premium sound system.

The very last XK (X-100) rolled off the Coventry production line in May 2007, the final production figures being:

XK8 coupe	19,748
XK8 convertible	46,760
XKR coupe	9,661
XKR convertible	13,895
TOTAL	90,064

2006: Going Down Alloy Alley

In January 2005, Jaguar announced their Advanced Lightweight Coupe (ALC) at the NAIS Motor Show in Detroit, and it was obvious that this would form the basis of a new XK. Jaguar's XK competed in two major market sectors, Grand Tourers (with a high degree of luxury) and Sports Cars (with high performance and good handling), and these sectors had increased in size dramatically since the 1990s. Jaguar's replacement (to be coded X-150) therefore had to target all these areas, indeed, a wider audience than the outgoing X-100 served.

Initial decisions included building the new car from aluminum (a material Jaguar was now geared up for with the introduction of the XJ in 2003). Next the styling had to be more modern; yet retain much of the essence of the old car, which had been so successful. Thirdly, integrity; all previous cars had been designed as coupes and then adapted for convertibles with extra strengthening and compromises. In this case, the X-150 was designed as a convertible with built-in integrity and the coupe derived from it. CAD (computer-aided design) sped up the whole process to bring the new car to market much quicker than previously.

More aggressive styling was adopted with classic ground-hugging proportions, a long bonnet, and minimal body overhangs. Sitting on a longer wheelbase from the front, it retained a mouth, though larger, with a deeper undervalance, and more substantial bumper still incorporating the auxiliary lighting. The bonnet incorporated a more prominent power bulge and the windscreen was more steeply inclined.

From the side, Jaguar incorporated what would become their signature power vents in the front wings. Particularly with the coupe model, the short overhangs, the wheels filling the wheel arches, and the pronounced haunch in the rear wings made for an impressive style.

At the rear of the coupe there was no separate boot lid like the previous model; instead, a hatchback hinged at the top provided excellent access to the large boot area. The convertible retained a lid. Overall rear styling was more impactful with modern lighting and a substantial bumper arrangement.

Interior design was all new with a mix of modern and traditional materials giving a choice of trim to suit individual fashions. There was 59 millimeters more seat travel, 54 millimeters more legroom, 31 millimeters more headroom, and 32 millimeters greater shoulder room. New seating still incorporated the 2+2 arrangement, still with minimal rear legroom for larger passengers. The new dashboard layout incorporated for the first time a touch-screen to operate various in-car functions, with a built-in satellite navigation system. The instrument cluster in front of the driver was cleaner with a centrally mounted vehicle information readout and a simple bar gauge for fuel content. Switchgear was reduced and more logical in operation and there was a new three-spoke steering wheel. There was also a dual-zone air conditioning system and a choice of audio systems and Bluetooth connectivity.

This was the first Jaguar to feature Smart Key operation with keyless ignition via a prominent red starter button on the center console, and keyless entry to the car.

With the existing 4.2-liter V-8 engine satisfying current legislation demands and proving to be a better performer in the lighter all-aluminum body, it was retained but with new multihole injectors improving fuel spray pattern in the combustion chambers. There was also a new fully electronic throttle control. Jaguar felt there was a need to improve the sound effect of the engine, so a semi-active exhaust system varied the flow of gases through the main silencer box, depending on pressure. Even acoustically tuned tailpipes were fitted to reduce boom, and an underfloor resonator balanced the sound generated from each cylinder bank of the engine.

An uprated version of the existing six-speed ZF automatic transmission incorporated a Bosch Mechatronic shift—an electro-hydraulic shift mechanism with adaptive strategy to respond to both road conditions and driving style. "J-gate" gear selection was retained with their new Sequential Shift, a fully automated drive mode adaptable to individual driving styles, accompanied by paddle controls on the steering wheel.

The lightweight structure of the new car allowed for the best distribution of suspension and axle components, the car retaining Jaguar's proven combination of unequal length wishbones at the front, and wishbones using the driveshafts as upper links at the rear. A new version of the CATS suspension included a two-stage adaptive damping system. Steering retained the Servotronic 2 system, reengineered with a faster steering rack.

The brake system was also upgraded with larger discs, brake assist, and a four channel ABS system along with an electronic park brake function.

Safety features were vastly improved with various onboard systems. A dynamic headrest system guarded against

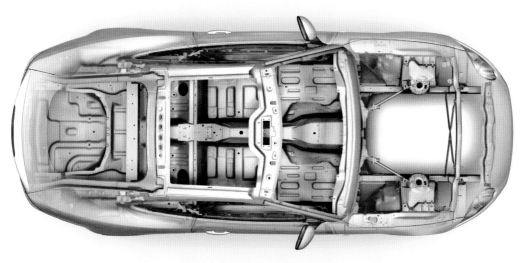

(ABOVE) The XK under the New XK, steel versus aluminum. The significantly redesigned New XK still displays a considerable similarity to the original car.

(LEFT) All-aluminum construction was a major change Jaguar made in 2003 for the XJ saloon. The experience was carried forward for the New XK, achieving lightness and rigidity plus better recyclability.

(TOP) The steering, suspension, and braking systems were uprated for the New XK.

(BOTTOM) Designed as a convertible first, the coupe benefited from the rigidity built into the new car.

whiplash injuries by automatically pushing the seat headrest forward to support the head in an accident. Intelligent front-seat airbags would sense seat positions for appropriate deployment on impact. There was a unique Pedestrian Safety System that, upon the car's impact with a pedestrian at the front, would deploy an explosive charge to lift the bonnet from the rear to protect the pedestrian from riding up the bonnet and smashing into the windscreen. There was a state-of-the-art Thatcham security system, dynamic and traction control, and available adaptive cruise control. Finally, for the more vulnerable convertible model, two hidden aluminum hoops were built into a reinforced structure behind the rear seating that would deploy in the event of a rollover. Activated via a solid-state gyro sensor,

the hoops protected the occupants from being trapped under the car.

The public launch of the New XK, as it was called, took place in September 2005 for 2006 release. The car received numerous accolades like the Car of the Year and Luxury Car of the Year.

Nearly 12,000 New XKs were delivered to customers in 2006, close to double the number of the previous XKs sold in its first full year of production.

2007: The "R" Returns

Initially, there was no XKR (supercharged) New XK, but only months after the introduction of the X-150, Jaguar announced the XKR model equivalents.

Using the same power unit as before, twin air inlets and variable inlet camshaft timing continually adjusted the

The New XKR arrived just a few months after the XK with subtle external differences to the normally aspirated model.

timing on both head banks, making for improved torque figures, particularly at lower revs. This, combined with the significantly stiffer and lighter aluminum body, resulted in a substantial increase in performance over the old model, with a power-to-weight-ratio leap of 12 percent. The transmission software was updated to match increased engine performance.

The New XK had been praised for its refined and sporting exhaust note. For the XKR, this was enhanced further, with a four-pipe system.

Several minor revisions served the extra performance. Spring rates were increased by 38 percent at the front and 24 percent at the rear, with a rear-suspension brace to manage this increase. The Servotronic steering system was retuned and front brake discs were enlarged from 326 millimeters to 355 millimeters with an extra 2 millimeters of thickness.

Trim changes to differentiate the XKR from the XK included a new front bumper with color-coordinated trim, different alloy mesh for the grilles and louvers in the bonnet. These front-wing power vents were finished in alloy instead of body color and new alloy wheels incorporated the "R" logo in their center caps.

Unique sports-style seating was adopted with better lateral support. The "R" logo featured prominently, and there was a unique style of alloy trim.

In March 2007, Jaguar introduced a new XKR limited edition model, the Portfolio. The car was fitted with 20-inch Cremona five-spoke alloy wheels, polished aluminum power vents with indicator side repeater lights, and was finished in just one color, Celestial Black (or Liquid Silver for UK and Swiss markets). Alcan created the most powerful braking system on a Jaguar at the time, with 44-millimeter-larger front discs with race-bred crescent-shaped grooves and six-pot calipers. Internally, the Portfolio benefited from an engine-spun aluminum trim, contrast stitching to the soft grain leather, "Portfolio" legend treadplates, and a Bowers & Wilkins hi-fi system.

2008: Faster and Faster

Yet another limited-edition model appeared at the 2008 Geneva Motor Show, the XKR-S, only available as a coupe and for European markets only. The engine-management system was recalibrated, providing 420 brake horsepower and 413 pounds-feet of torque, reducing the 0- to 60-miles-per-hour acceleration time to 5.2 seconds. This model incorporated the Alcan braking system from the Portfolio, but also new springs, anti-roll bars, and shock absorbers and a faster ratio steering rack.

Styling changes included sill extensions and rear spoiler, available only in Ultimate Black paint. Inside, Piano Black veneer with twin-needle stitching to the soft grain leather, and a luxury headliner, separated this model from others.

Mid-year in 2008, Jaguar celebrated the 60th anniversary of the postwar XK120 sports with a revised XK model to complement the XKR and XKR-S. Called the XK60, it was fitted with standard 20-inch alloy wheels, a new front spoiler and rear valance panel, chromed power vents, and grille mesh.

2009: Generation III Is Born

On January 12, 2009, Jaguar announced major powertrain changes to their entire car range, offering more power and greater efficiency. For the XK range this boost was the AJ-V-8 Gen III 5.0-liter engine in both normally aspirated and supercharged forms.

Jaguar's most advanced engines to date, they utilized a stiff high pressure cast-aluminum block with cast-in steel liners, aluminum cylinder heads with four valves per cylinder, and spheroidal-graphite cast iron crankshafts and forged-steel connecting rods. These direct-injection gasoline engines were more compact. The sixth-generation twin vortex system supercharger was fitted to the XKR version. A high helix rotor design improved supercharger thermodynamic efficiency by 16 percent over the old 4.2-liter engine.

A key feature of the new engines was the uprated, centrally mounted, multihole fuel injection system. There was also a new variable camshaft timing system triggered by positive and negative torques generated by opening and clos-ing the intake and exhaust valves. The normally aspirated engine had camshaft profile switching on the inlet camshaft.

An innovative reverse-flow cooling system delivered thermodynamic and friction improvements, pumping coolant through the cylinder heads before it flowed through the block and returned to the radiator. The XKs exhaust note was further enhanced and, for the XKR, Jaguar engineered an intake feedback system with intake manifold pressure pulsations fed into an acoustic filter at the rear of the engine.

Jaguar introduced their JaguarDrive transmission controls, replacing the "J-gate." On start-up, a rotary control rose from the center console for gear selection; steering wheel paddles remained for more direct selection. The gearbox was an uprated XF six-speed 6HP28 unit.

Another first for Jaguar was the Active Differential control as a final part of the supercharged driveline. Designed to give improved traction and stability, the electrically controlled differential continually adapted to a driver's demands, operated by an internal electric motor. The differential contained a multiplate clutch transmitting or vectoring torque to the wheel to ensure grip.

The latest version of Jaguar's Adaptive Dynamics replaced the CATS suspension system, which controlled body vertical movement, roll rate, and pitch.

There were a number of significant styling changes to the 5.0-liter models. Externally, a new frontal design meant new panels, reshaped nose, chromed detailing to grilles, new lighting, and vertical mesh/chrome cooling ducts, and at the side redesigned power vents (again!). There was new rear bumper and slight trim changes, with color differences according to model.

Inside higher-end models a suede-cloth headlining was fitted, a new center

(TOP) The first facelift XK/XKR with its modernized frontal aspects.

(LEFT) Under the bonnet, an entirely new 5.0-liter V-8 engine.

(BOTTOM) The much-enhanced interior of the 5.0-liter models, in this case an XKR. Note the new JaguarDrive control on the center console for gear selection.

The fast XKR-S, offering another revised front and many other aerodynamic touches, arguably the fastest of the road-going XKs.

console incorporated the "JaguarDriver" control, steering wheel leather wrapping was used, and there were minor styling changes to instruments and switchgear. Heated and cooled seating was now standard on top-of-the-line models with extra seat adjustment; door trims were amended, and a better finish overall was applied to the whole interior.

2010: 75 and 175

In 2010, Jaguar introduced a limited run of models to celebrate its 75th anniversary. The XKR 75 models were finished in Stratus Gray with optional gray striping along the top of the front and rear wings, plus new sill extensions, a larger rear spoiler, and 20-inch Vortex alloy wheels. Inside there was Piano Black veneer with XKR 75 treadplates and a dashboard card signed by Jaguar's Design Director Ian Callum and Chief Test Engineer, Mike Cross.

Performance was improved with the engine developing 530 brake horse-power, top speed increased to 174 miles per hour with a revised ECU, upgraded torque converter, along with suspension, brakes, and steering retuning.

A version was produced for the US market, the XKR 175, but with fewer performance upgrades.

2011: Yet More Changes and Models

At the 2011 Geneva Motor Show, Jaguar announced subtle changes to the XK range. The latest headlight technology was adopted, incorporating LED signal and running lights. The light units were of an entirely new design, necessitating changes to the front and side panels; this was combined with a larger grille

A more aggressive look to the XKR-S from any angle, and this was probably the common view for many drivers of other cars!

and new bumper design and cooling vents. Because of the changes to the front wings, new horizontal power vents were fitted. The boot lid incorporated a slimmer chromed finisher, allowing room to fit a traditional Jaguar chromed leaper. Paint finishes were revised and expanded, and slight trim changes were made according to model.

The interior was also refreshed with revised switchgear, ambient lighting, and new steering wheel design; performance front seating with significantly more lateral support was available for the Portfolio and XKR models.

The biggest news of 2011 was a new, higher performance production model, the XKR-S (a named used much earlier for a very limited-production model). The most expensive XK model, initially it was available only as a coupe; its XKR engine was remapped to boost performance to 550 brake horsepower, which in conjunction with transmission modifications produced a 0- to 60-miles-per-hour time of just 4.2 seconds, with a limited top speed of 186 miles per hour!

The front suspension was considerably revised with a fully machined steering knuckle, increasing camber and caster

stiffness. The rear suspension geometry was revised with rear wheel steer optimized for maximum agility, while spring rates were increased by 28 percent. Ride height was reduced. The Active Differential system was also reprogrammed to match the higher speeds.

Huge 380-millimeter front brake discs with 376-millimeter discs at the rear were combined with aluminum calipers with pad area increases of 44 percent and 31 percent to provide outstanding brake performance.

An extensive styling package was fitted for the XKR-S. A wider front

The interior of the XKR-S with the performance seating and revised trim.

bumper arrangement incorporated a bigger air intake, a carbon-fiber splitter bar, and twin side nacelles. At the side of the bumper, the vertical panels channeled air down the sides of the car, smoothing out airflow. A unique rear wing incorporating a carbon-fiber inlay worked in conjunction with a rear apron with a carbon-fiber diffuser, reducing overall lift by 26 percent.

The exterior trim was in gloss black finish. Dark 20-inch Vulcan alloy wheels framed brake calipers finished in a choice of gunmetal or red.

Internally, the XKR-S featured performance seating with carbon-leather soft grained trim with contrasting stitching.

Jaguar also offered a special Carbon Fibre pack for this model, with carbon-fiber bonnet louvers, carbon-fiber power vents, door mirror covers, and boot lid finisher.

In November, Jaguar finally announced a convertible version of the XKR-S. UK prices for the XKR-S models were nearly £20,000 higher than equivalent XKR models.

2012: Yet More . . . !

In 2012, Jaguar kept the XK in the public's focus with another new model, the SE (or Artisan as it was called in the brochures). The SE offered performance seating, the best Scraffito grain leather upholstery, contrast micropiping and stitching in a choice of navy or truffle, along with special dark walnut veneer or dark aluminum. The changes also included 20-inch Orona alloys and Bowers & Wilkins hi-fi system.

2013: Ultimate Performer

In 2013, Jaguar announced their most ambitious XK of all, the XKR-S GT, developed by Jaguar's Engineered to Order Division as the ultimate track-focused Jaguar.

A range of specially produced carbon-fiber components were fitted to

(ABOVE) The ultimate ground-hugging XKR-S GT, a road car designed and built for track use.

(LEFT) Featuring extensive use of aerodynamics, carbon fiber, and many mechanical components from the then new F-TYPE, the XKR-S GT was the fastest and rarest of the XK models.

The last of the line normally aspirated XK, the Signature model.

the car, like an extended front splitter, dive planes, and elevated rear wing to improve aerodynamics and downforce. Finished in Polaris White (with a few examples in other colors by individual request), the body incorporated unique graphics. The black interior featured the performance seating in a mix of leather and suedecloth incorporating XKR-S GT logos, a suedecloth headlining and aluminum paddle shifters and pedals.

The supercharged engine producing 550 brake horsepower was taken from the XKR-S. The new F-TYPE sports car contributed front- and rear-suspension uprights, wheel bearings, bushings and rear subframe, increasing lateral suspension stiffness. There was a new spring and shock absorber design, a motorsport derived twin-spring system mated to Jaguar's Adaptive Damping. Spring rates were substantially stiffer than the XKR-S—by 68 percent in front and 25 percent at the rear.

Steering was taken from the F-TYPE with a faster ratio, while the XKR-S GT was the first production Jaguar to be fitted with a carbon-fiber braking system. The ventilated and cross-drilled lightweight discs were 398 millimeters at the front and 380 millimeters at the rear, grabbed by six-pot monobloc calipers front and four-pot at the rear. All other aspects of the braking system were also upgraded.

The car was fitted with unique 20-inch alloy wheels with specially developed Pirelli Corsa 255/35 and 305/30 front/rear tires.

2014: The Grand Finale

For the last year of XK production, Jaguar rationalized the range, and introduced two new models, the Signature and Dynamic R running alongside the XKR-S.

The luxury-focused Signature featured the normally-aspirated 5.0-liter engine, plus a host of extras, including 20-inch wheels, a reverse-park camera, luxury seating, ebony veneer, and bright metal pedals.

The Dynamic R was the supercharged model, fitted with a combination of Jaguar's Black, and Speed packages (see sidebar area). It also sported new diamond sewn performance seating and soft-grained leather.

For the US market the "Final Fifty" was the limited edition, finished in Ultimate Black with Ivory interiors. The car included most of the Dynamic R's features plus a commemorative plaque signed by Ian Callum.

On July 24, 2014, the very last XK left Jaguar's production lines after a production run of 54,549 cars, making the entire XK range from 1996 to 2014 the most successful ever of all Jaguar's sporting models.

More Specials

Jaguar offered several packages and limited options throughout the life of the X150. As early as 2007, they produced a special "150" styling kit with a new front and rear valance, upper and lower mesh grilles and sill extensions plus new rear lower bumper and square exhaust tailgates. The kit came in primer finish to be fitted by Jaguar dealers.

In 2011, Jaguar introduced "in-production" additional packages that buyers could choose.

The XKR Speed Pack offered engine and suspension retuning to achieve a new maximum speed of 174 miles per hour. Available only on coupe models and in a range of seven exterior paint finishes, it included body-colored side skirts and rear diffuser, chrome finish to the window surrounds, mesh grilles, side vents, boot finisher, and red brake calipers with 20-inch Kasuga alloy wheels.

The XKR Black Pack offered more visual impact, again for coupe models.

The last of the line supercharged Dynamic R.

Black finish was applied to window surrounds, grilles, and power vents, complementing black alloy wheels, body-colored spoilers, and a sweeping XKR side graphic.

The XKR Dynamic Pack provided an enhanced version of the Speed Pack, plus features from the XKR-S model.

For the Indian market, there was a limited run of XKR Special Editions in coupe and convertible forms; this was effectively a lower-production run of the SE (Artisan) model.

In 2014, Jaguar introduced the XK66 primarily for the German market. Based on the Portfolio model and available in only two colors, British Racing Green or Stratus Gray, the "66" referred to the number of years since the XK120 debuted. Special badging was included.

SPECIFICATIONS

MODEL	4.0-liter XK8	4.0-liter XKR	4.2-liter XK8	4.2-liter XKR	4.2-liter XK
ENGINE SIZE	3,996cc	3,996cc	4,196cc	4,196cc	4,196cc
CARBURETION	Fuel Injection	Fuel Injection	Fuel Injection	Fuel Injection	Fuel Injection
MAXIMUM BHP	290@6,100	370@6,150	300@6,000	400@6,100	300@6,000
MAXIMUM TORQUE	290@4,250	387@3,600	310@4,100	408@3,500	310@4,100
AUTOMATIC	5-speed	5-speed	6-speed	6-speed	6-speed
0 TO 60 MPH	6.5 sec.	5.2 sec.	6.1 sec.	5.2 sec.	5.9 sec.
TOP SPEED	155 mph	155 mph	155 mph	155 mph	155 mph
AVERAGE FUEL CONSUMPTION	23 mpg	23 mpg	24.9 mpg	22.9 mpg	25 mpg

MODEL	4.2-liter XK4 (X-150)	5.0-liter XK	5.0-liter XKR	5.0-liter XKR-S	5.0-liter XK-RS GT
ENGINE SIZE	4,196cc	5,000cc	5,000cc	5,000cc	5,000cc
CARBURETION	Fuel Injection	Fuel Injection	Fuel Injection	Fuel Injection	Fuel Injection
MAXIMUM BHP	420@6,250	385@6,500	510@6,500	542@6,000	550@6,500
MAXIMUM TORQUE	413@4,000	385@6,500	510@6,500	502@2,500	502@2,500
AUTOMATIC	6-speed	6-speed	6-speed	6-speed	6-speed
0 TO 60 MPH	4.9 sec.	5.2 sec.	4.6 sec.	4.2 sec.	3.9 sec.
TOP SPEED	155 mph	155 mph	155 mph	186 mph	186 mph
AVERAGE FUEL CONSUMPTION	22 mpg	25 mpg	23 mpg	21 mpg	23 mpg

CHAPTER ELEVEN

NEW MARKETS
IN THE MILLENNIUM

After the demise of the Mark 2 compact saloon in the late 1960s, Jaguar concentrated on their top-end luxury model, the XJ. Those cars served them well for many years but allowed other manufacturers like BMW and Mercedes to seize what would turn out to be an increasingly important market with their smaller prestige models like the 5 Series and E-Class.

With financial and technical support from Ford, Jaguar launched the S-Type (code X-200) medium-sized saloon in 1999. So successful was the car, that in 2001 a further new model, the X-Type, appeared to challenge even smaller prestige saloons like the BMW 3 Series. Jaguar's first venture into smaller cars since the 1.5-liter SS Jaguar saloons from the 1930s, this move also proved successful, giving Jaguar a four-model line-up at the turn of the century (X-Type, S-Type, XJ, and XK).

With a need to grasp modern technology and ensure the continued future of the big luxury saloon, Jaguar made great strides forward when the new XJ (coded X-350) was launched in 2003. This completed the make-up of models

for the millennium seeing the business through another interesting period and to yet another new owner, Tata of India.

1998–2000: The S-type Reborn

It was at the British Motor Show in 1998 that the name Jaguar S-Type was seen for the first time in thirty years, reestablishing a medium-sized saloon in a market that the company had dominated in the 1950s and 1960s. Jaguar aimed the new car at a younger male and female market and fleet buyers who had previously not purchased an XJ due to its size and price.

The S-Type was virtually an entirely new car from Jaguar's perspective,

although it retained the AJ-V-8 4.0-liter engine first seen in the 1996 XK and 1997 XJ.

The foundation for the new car was a floorpan jointly developed with Ford for both the S-Type and US Lincoln LS. It was built in the States and shipped over to Jaguar's factory along with other structural pressings. Other specific body components were produced in the UK to meet specific S-Type needs. Many new and innovative design features were built into the car like the limousine-style doors, where the top edge extended into the roof for ease of entry/exit and to eliminate wind noise. The side of the body was formed in a single piece from the A-post to the rear wing (over 3.5 meters in length) for a stable structure

The Jaguar S-Type, first return to the medium-sized luxury saloon market. An all new design separated it from Jaguar's traditional saloon styling in the XJ.

and ease of manufacture. Special steel that hardened with the paint process was used for vulnerable areas like the bonnet, door skins, and front wings. A one-piece structured pressing was used for the bonnet to provide the many curves required by the styling. The front end was built as a single piece in plastic for safety, and plastic was also used for the sill covers. Hydroform technology was used to inject oil into the steel tubes of items like the radiator cross-member. Eighty-five percent of the body was double-sided zinc coated. The car had triple door seals and computerized body assembly; laser alignment ensured an excellent ft and finish. All the body panels were made for ease of access and replacement with smaller service panels

available to avoid the need for full panels during a repair.

The external styling of the car was pure Jaguar, designed by their then Chief Styling Director Geoff Lawson. The backward-sloping frontal aspect incorporated a radiator grille resembling both the Mark 2 from the 1960s and the C-Type sports/racing car of the 1950s, flanked by quad elliptical headlamps with impact resistant poly-carbonate lenses. The substantial front bumper incorporated an air intake beneath with space for foglamps; on top were twin chromed "blade" finishers. The beautifully sculptured bonnet panel, hinged at the rear, incorporated a growler badge and led to a large raked windscreen.

The large glass area to the doors and screen aided visibility and added light to the interior of the car. The sides of the car incorporated two swage lines and rubbing strips and the body swept down to the rear wing area. The rear provided a little hint to previous Jaguar styling with a matching curved bumper with chrome blades, a large boot lid, and wraparound lighting.

Internally, it was all new again, certainly with a more youthful approach to design available with leather and wood veneer, with the option of cloth upholstery. Split rear seating providing a 60/40 fold-down arrangement for more luggage accommodation was a Jaguar first. The dashboard was ergonomically designed with a facia-mounted ignition/

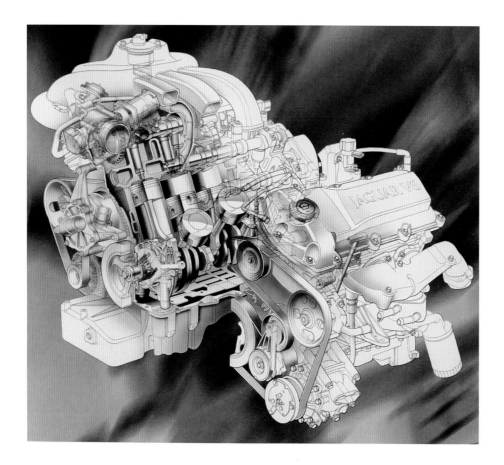

starter and instrumentation in front of the driver. The half-moon-shaped center console incorporated myriad features according to model, including audio controls, cassette storage, an integrated phone system (where fitted), air conditioning (standard on all models), and a five-function computer with an eleven-language message center display. The air conditioning providing dual-zone operation was jointly developed with Ford. A satellite navigational system was

(LEFT) The start of a range of V-6 engines initially in the S-Type, but eventually leading through to all saloon models into the millennium.

(BELOW) The interior of the early S-Type was a culture shock for many Jaguar traditionalists, governed by certain aspects of the Ford parts bin.

also available which, for the first time, incorporated voice activation.

Along with the 4.0-liter V-8, there was a 3.0-liter V-6 developed by Jaguar but shared with the US Lincoln. For this engine, a rigid 60-degree twist-forged steel crankshaft with fully machined balance weights, undercut rolled fillets, and heat-treated pins/journals was used. Camshafts used a low-profile Morse silent dual-chain drive and hydraulic tensioner. Cast-in thin-wall iron cylinder head liners, low-pressure cast heat-treated 319-grade aluminum cylinder heads, cast iron, light silicon/molybdenum exhaust manifolds and directly mounted engine ancillaries completed the specification for the new engine. The 3.0-liter developed 240 brake horsepower at 6,800 rpm.

Two gearboxes were fitted to the early S-Types—a five-speed Ford automatic transmission incorporating Jaguar's "J-gate" and a Getrag Type 221 five-speed manual gearbox operated through a 240-millimeter diameter self-adjusting clutch.

A bespoke powertrain electronic controller managed primary engine and transmission functions and secondary systems.

A plastic deformable fuel tank was used for the first time, mounted below the rear seat pan.

Front suspension was all independent, with twin unequal length wishbones and coil springs with telescopic shock absorbers. The wishbones were lightweight alloy forgings, the uppers having integrated ball joint, mounted directly to the body. Coil springs were coaxial with the shock absorbers mounted between the lower wishbone and body.

There were two crossbeams. One, a steel fabrication, provided the location for the lower wishbone mounting, anti-roll bar, and engine cooling system; the other (positioned behind the first), was an alloy casting providing

the mounting for the rear arms of the lower wishbone, the power steering rack, and engine mounts. An unusual "swan neck" aluminum design front knuckle casting with steering arm provided location for the upper and lower ball joints, disc shields, brake calipers, and wheel bearings with integral wheel speed sensors.

The independent rear suspension was also new. Unequal length twin aluminum wishbones had constant velocity jointed driveshafts. Coil springs, coaxial with shock absorbers, were mounted between the lower wishbone and the body. These were heavily inclined in plan and side views to provide good anti-squat and anti-lift characteristics. The wishbones were mounted on a crossbeam, isolated from the body by bushes, with the steel crossbeam also providing the mounting for the differential and anti-roll bar. Jaguar's CATS suspension was available for the S-Type as an option.

The steering had electronic control, a speed proportional design with a variable-ratio rack. The braking system incorporated large-diameter ventilated discs (300-millimeter front with twin pistons, 288-millimeter rear with a single piston), with a four-channel anti-lock system incorporating electronic brake distribution and traction control. A conventional handbrake arrangement was used.

Standard equipment wheels were 16-inch, with 17-inch as an option, both with Pirelli tires.

The S-Type was launched in a choice of just three models, the 3.0 liter, 3.0-liter SE (Special Equipment), and 4.0 liter.

Over the first 12 months of production, various color changes took place to match similar changes in the other Jaguar models, and a BBS Monaco 18-inch split-rim alloy became available, part of the new R-Performance options range.

Sales for 1999 surpassed their 75,000 figure from ten years previous. The

major contributing factor had been the S-Type with 39,000 sold from April to December, of which 77 percent were exported.

Announced towards the end of 2000 were some upgrades to the S-Type with the power steering following the same changes as other Jaguar models. A Sports Pack became available to upgrade the standard saloon with CATS suspension, 17-inch alloy wheels, and sports seating. An annoying aspect of the early cars was that if a CD auto-change was fitted, it was located in the passenger side glove box taking up most of the space. For the 2001 model year this was removed to the boot area. Under the boot floor a new module had been created to store the jack and tools, previously kept on top, and the side body rubbing strips were removed giving the car a cleaner line.

2001: X Marks the Spot

With the S-Type becoming Jaguar's most successful saloon ever, reaching nearly 107,000 sales to date, the company announced to the world another small new saloon (coded X-400), the X-Type. The goal was to attract an entirely new market to the brand, the under 40-year-olds, with 80 percent of anticipated sales coming from conquests taken from those who would have normally bought the likes of a BMW 3 Series.

It took a mere two years to launch the X-Type, a car to be built at an ex-Ford facility near Liverpool in which £300 million had been invested. Every aspect of production had been examined carefully to ensure quality and supply. Outside component suppliers were reduced from 350 with the S-Type to just 130 for the X-Type, and 90 percent of the car, by weight, was now recyclable. The body was reportedly 30 percent stiffer than any equivalent-sized saloon from other manufacturers.

(ABOVE) Jaguar's smaller saloon, the
X-Type, with obvious styling connections
to its larger brother, the XJ. Although
aimed at a new market, there were a
significant number of Jaguar owners who
took this opportunity to downsize.

(RIGHT) The interior of the X-Type, in
this instance a Sport model, with the dark
stained woodwork and a new dashboard/
console layout that would eventually find
its way into other models.

The new interior layout of the S-Type returned to more conventional Jaguar appearance and echoed that of the smaller X-Type.

Styling took cues from the larger XJ but incorporated aspects of the S-Type's roofline and boot area. Inside was the same, with touches of the XJ plus the S-Type's instrument pack and better-quality switchgear and trim. For the first time the complete dashboard was assembled as one unit. Top-of-the-line models got a telematics system incorporating 120-watt Alpine sound and a wide touch screen for finger-tip controls of audio, air conditioning, and satellite navigation, plus a TV and phone where fitted.

New seating with a choice of trims provided an enhanced feel of quality and there was more storage space than in other models. The boot area was also quite large.

The X-Type was launched with just two engines, the existing 3.0-liter V-6 (from the S-Type) and a new 2.5-liter V-6 version of it developing 194 brake horsepower. Transmission for either was the five-speed manual or Ford automatic, both from the S-Type.

The X-Type's biggest innovation was all-wheel-drive. Jaguar's Traction-4 full-time all-wheel-drive system allowed 60 percent of torque to be delivered to the rear wheels to provide the feel of conventional rear-wheel drive. If the system's sensors detected a speed difference between the front and rear road wheels, torque could be transferred via a viscous coupling to correct the imbalance.

The front suspension had two bearing top mounts and twin-tube McPherson struts with Hydrobushes for better compliance. The rear axle/suspension assembly was unique to this model, although in both front and rear situations, many components came from the Ford parts bin.

The X-Type was launched initially in a choice of six models, standard trim, SE (Special Equipment) and Sport, with the 2.5- or 3.0-liter engine. The Sport option included a discreet boot spoiler, 17-inch wheels, matte black window surrounds, color keyed bumper blades, grille, gray stained wood veneer, and

stiffer suspension settings. The Sport theme had been started on previous XJ models, commencing with the XJ40 and the X-Type followed this pattern to appeal to a younger market.

The same principle followed now with the S-Type, as a Sport version of that became available during 2001. Dechroming of exterior brightwork was the first noticeable difference, along with a mesh grille plus a more contemporary interior look with the dark stained woodwork and sportier seating. Wider wheels were standard.

By the end of 2001, S-Type sales for the year were down to 38,325, while the X-Type in its first year of production achieved sales of 55,616.

2002: New Power Units and a New Car

Much was afoot in 2002. Despite the initial success of the X-Type, it made no inroads on BMW's 3 Series market because the majority of those sales came at the entry level, below the X-Type's UK price of £20,000 (also a company car taxation limit). To rectify this, Jaguar introduced a new engine in all three forms—X-Type, X-Type Sport, and X-Type SE.

The engine was a 2,099cc, 24-valve, four-camshaft V-6 delivering 157 brake horsepower with torque of 148 pounds-feet for an estimated top speed of 130 miles per hour. Derived from the existing 2.5-liter V-6, it had the same bore but with a 66.8-millimeter stroke.

The 2.0-liter models were front-wheel drive. The drive was connected via equal-length half shafts, the geometry calculated to ensure they were perfectly aligned under torque. Special constant-velocity joints at the outward ends reduced friction, and unique strut top mounts were used to eliminate friction there under torque.

The announcement of the supercharged S-Type R meant a revised appearance for that model.

Retuned suspension matched the new engine and car weight. Standard equipment was the Getrag gearbox with revised ratios; a Jatco FPD five-speed automatic was also available.

The S-Type bodyshell was reengineered, reportedly now 10 percent stiffer through the use of laser-welded roof joints and high-strength steel in key areas and better-quality fit to incorporate new side-curtain airbags.

At the front, there was a new grille incorporating a growler badge and modified surround. Xenon lighting was now available as well. From the side, as with other models, the rubbing strips were removed, and there were new door mirrors and a wider choice of alloy wheels.

Due to some complaints about the interior, there was a complete change adopting the same style as the X-Type, which itself was similar to an XJ! This included the telematics system and other enhancements standardized from the X-Type, like uprated sound systems. There was also additional storage plus modified door panels and other trim.

To improve comfort, foot pedals could be adjusted electrically by a control in the driver's door. There were now no fewer than three seat combinations depending on model, all using magne-sium seat frames for lightness, and with electric seat adjustment.

Safety enhancements followed current Jaguar practice from the other models, including Adaptive Restraint Technology.

Mechanically there were new engines and transmission. These included the 2.5-liter V-6 used in the X-Type, yet retaining the S-Type's rear-wheel drive. The 3.0-liter engine came in for some minor revision to improve fuel economy by revising the three-stage variable geometry induction system and cam phasing.

The 4.0-liter engine had been upgraded to 4.2 liters and was now fitted to the S-Type. Adding the supercharged version created the S-Type R model, one of the fastest in its class. For use in the S-Type, supercharger speeds were increased by 5 percent.

Manual transmission remained unchanged, but the new six-speed Mercedes automatic was now fitted to V-8 S-Types.

S-Type suspension was now somewhat standardized with other models, particularly the X-Type, stiffened up to suit the S-Type R's extra performance requirements. An industry first at the time was an electronic parking brake for the automatic transmission models. A switch on the center console applied or released the handbrake mechanism, or it was done automatically upon removing the ignition or when accelerating from a standstill. The system worked by applying caliper levers through a parking brake actuator mounted on the rear subframe with an electronic control unit mounted in the boot. Larger brake discs were now fitted to S-Types incorporating the four-channel anti-lock system and panic-assist feature.

At the 2002 British Motor Show, Jaguar presented the special limited-edition X-Type Indianapolis, a model available only in three colors: Ebony Black, Platinum Silver, or Ultraviolet Blue with Alcantara black sports seating and special veneers. Sitting on 18-inch wheels, the model was available with either the

2.5-liter or 3.0-liter V-6 engines. Standard equipment included xenon headlights, air conditioning, and reverse park aid.

Although sales through 2002 had fallen somewhat for the S-Type, the overall figures were still excellent at 36,000 units, while the X-Type was at its peak, almost doubling S-Type sales to over 69,000.

2003: New XJ

In 2002, Jaguar introduced the New XJ (coded X-350), identifying it as "The Most Advanced Car Ever." To avoid too radical a style departure, the design featured a somewhat reinvigorated body with a lot of elements relating to the much smaller X-Type but all the better for the larger dimensions. Larger overall than any previous XJ, the car had more room inside and in the boot area.

From the front the established four-headlight styling of all XJs was continued with more modern lighting, another revised grille (individually designed according to model), the bumper bar of the current "corporate" style with chromed blades and integrated auxiliary lighting. The bonnet was marginally shorter, wheels were positioned closer to the extremities of the body and filled the wheel arches better. The rear area was smartened up, yet still easily identifiable as an XJ.

Despite similarities to what had gone before, the X-350's styling hid Jaguar's new all-aluminum bodywork with lighter magnesium crossbeam along the bulkhead and for the interior seat frames. The car also featured an industry-first rivet-bonded joining of panels using self-piercing rivets combined with aerospace-sourced epoxy adhesives to join the aluminum pressings and castings. The body was also built from fewer panels than previously, producing a car that was not only 60 percent stiffer but also 40 percent lighter.

The car initially came with a choice of four engines, the existing 3.0 liter V-6 (reinstating the XJ6 name), the 4.2-liter V-8, and the 4.2-liter supercharged V-8, plus a new 3.5-liter V-8 (derived from the 4.2) providing 262 brake horsepower, all mated to Jaguar's XF six-speed automatic transmission.

An entirely new suspension system brought the X-350 into a new realm of ride, comfort, and dynamic quality. The mechanicals were based on the S-Type and incorporated a self-leveling air suspension system. This ensured that full suspension travel was always available by increasing the spring's stiffness relative to payload. It also ensured that the car had consistent ride height and the ability to adapt to changing circumstances. A new version of Jaguar's CATS suspension was also fitted to all New XJs.

The interior of the cars was also new, following the design influences from the X-Type and later S-Types, along with quality switchgear, ventilation outlets, and Piano Black finished trim. Head, shoulder, and leg room were all improved and there was an extensive range of standard features as befitting the top-of-the-range models. Voice activation, electrically adjustable pedals, electrically controlled rear screen blind, and heated front screen were just a few of the features now available.

There were lots of first-time features for the XJ including radio signal pick-up embedded in the rear screen (no separate aerial), heated elements in the base of the front screen to prevent the wipers from freezing up in extreme conditions, and ground illumination that used the key fob to trigger lights under the door mirrors to assist in approaching the car in the dark. There was even a rear compartment console to control the audio/TV/video systems when fitted, with plug-in facilities for games consoles and headphones, and a four-zone air conditioning system.

(**ABOVE**) The New XJ at launch in 2003, in many ways looking like the model it replaced, but larger, with improved build quality, finish, and fitment. This is the entry model 3.0-liter XJ6, the first time that insignia had been used since the demise of the AJ16-engined model in 1997.

(**RIGHT**) With the introduction of new, smaller engines for the X-Type, an economical 2.0-liter diesel (from Ford's stable) would be installed, unusually in the transverse position with front-wheel drive only.

XJ ALUMINUM CONSTRUCTION

The all-aluminum construction of the X-350 would make for a lighter, more efficient, and environmentally friendly XJ, the technology eventually finding its way into all Jaguar models.

At Jaguar's Castle Bromwich factory, it took 30 percent less time to produce an X-350 body compared to earlier XJs. There, a total of 9 presses, including 5 giant Schuler's presses, formed the 125 aluminum components making up the bodyshell. In addition, 88 robots, linked by a sophisticated ethernet control system, applied adhesives and installed the self-piercing rivets in the XJ's structure. One hundred-forty laser measurements were taken upon completion of the body in white to ensure accurate assembly.

Each XJ body when completed was dipped in a bath of electrically deposited corrosion protection covering the entire surface, to maximize durability. Then the bodies received two coats of primer (the second matched to body color), plus two coats of water-based color basecoat followed by a further two coats of scratch-resistant clear coat. Each paint layer was baked for 30 minutes at up to 185 degrees Celsius (365 degrees Fahrenheit).

Additional innovations included a smart charging system to manage the battery load, cutting nonessential power if the battery was not at a reasonable charge, and a digital data bus fiber-optic communications system delivering high-speed information to relevant systems.

At launch, the range consisted of the XJ6 (3.0-liter V-6), the Sport (V-6 or 4.2-liter V-8), the SE—Special Equipment (all three normally aspirated engines), XJR (supercharged 4.2-liter V-8), and the Super V-8 (supercharged 4.2-liter V-8). There was only one (short) wheelbase bodyshell available and no Daimler-badged edition.

Returning to Jaguar's successful X-Type, another first in 2003 was the introduction of Jaguar's first diesel engine, a four-cylinder 1,998cc unit tracing back to the Ford TDCi unit. The engine developed 128 brake horsepower at 3,800 rpm with 243 pounds-feet of torque at just 1,800 rpm. The engine was lightweight with a stiff cast iron block with an aluminum ladder frame and aluminum cylinder head. Fuel injection was based on a high-pressure common rail direct system with variable geometry turbocharger.

Installation of the new engine in the X-Type was transverse, transmitting power through a standard X-Type five-speed manual gearbox to the front wheels. Significant sound deadening was added to the X-Type's bodyshell to eliminate noise, and changes were made to spring rates to match the altered weight of the power unit.

By the end of 2003, S-Type sales had amounted to 32,000 cars; X-Types were double that at 61,600 and the New XJ had sold a very healthy total of 30,800!

The facelifted S-Type from the rear with its cleaner lines, revised bumper treatment, and cleaner sides.

2004: "Extending" the Range

Jaguar had a busy 2004, starting with a significant facelift for the S-Type. At the front was a new look with a simpler but more assertive bumper design that varied according to model. Another new radiator grille appeared, set lower and wider, and the bonnet was made from aluminum with subtle changes to the curvature.

From the sideview, the sills had been resculpted to blend better into the overall styling of the car. The bumper bar wraparounds were also altered with a different depth to the "blades." The rear wing line, upper and lower bumper surfaces, and boot lid were all reshaped.

More significant changes took place at the rear with less curvature to many panels; chrome finishers and locks were changed, and the modern light units now wrapped over the tops of the wings. Bumper bar treatment was simplified and badging changed.

The changes improved aerodynamics and gave the S-Type a slightly more aggressive look as well as a more polished style.

Internally, although the general layout remained unaltered, the instrument array changed with clearer dials, and two message centers provided more information. Four trim options were now available with a new alloy finish or Bronze Madrona, Grey Bird's Eye Maple, or Burr Walnut veneer. There was also a Luxury Pack available with soft-grain leather, chrome highlights, a matching wood/leather steering wheel, and heated seats. Seat upholstery range was extended with a choice of cloth, conventional or perforated, or R-leather, with the option of dual-colored trim. Equipment levels were also enhanced.

With an expanding market for diesel cars, Jaguar had to address it, so for 2004 the S-Type became available with

(**ABOVE**) The first production estate car from the company, based on the X-Type, was a very attractive and uncompromised design that was received well by the public.

(**LEFT**) The 2.7-liter V-6 twin-turbo-charged diesel engine, a collaboration between Jaguar, Ford, and Peugeot-Citroen, was the most powerful diesel offered by Jaguar until it was replaced by an entirely new 3.0-liter engine.

CONCEPT EIGHT

Jaguar's Advanced Design Studio came up with a one-off concept based on the long wheelbase XJ bodyshell. The car was hand-finished and epitomized the ultimate in luxury.

Externally, the roof panel was replaced by full length/width darkened glass. This was the first XJ to have front wing "power vents." Paintwork was Purple Haze, a one-off blend of cherry and black. At the rear, there were larger chromed exhaust tailpipes.

Internally, that glass roof panel had strips of red LED lighting around the edges. The seats were entirely new, of quite modest proportions, with a split arrangement at the rear giving the effect of two separate seats. A full-length center console extended between the rear seats all the way forward to meet a conventional XJ console. Wood-work was smooth matte finished American walnut.

Jaguar's Concept Eight, a one-off but from which certain features would find their way into a limited-production Portfolio model.

a new 2.7-liter engine jointly developed with Ford and Peugeot-Citroen. The new block was made from compacted graphite iron. A four-camshaft unit with twin turbochargers and 24 valves, it produced over 200 brake horsepower with more torque than Jaguar's 4.2-liter V-8. Built for quiet running, it was also 40 percent more fuel efficient and only 15-kilograms heavier than an equivalent gasoline unit.

Installation of the diesel engine in the S-Type involved significant sound dead-ening work, like a double-skinned sump and an elastomeric-isolated composite cam cover. The suspension underwent changes with revised spring rates and

shock absorbers, low-friction ball joints, and the steering control was recalibrated. The engine was built at Ford's Dagen-ham Diesel Center in the UK.

The new engine was mated to an adapted version of the six-speed auto-matic transmission. A diesel-equipped S-Type was marginally cheaper to buy than an equivalent gasoline version.

For the X-Type came not a facelift but another new model, the first Jaguar production estate car. The market sector for cars of this size indicated that 18 percent of all new models sold were estate cars, so it was an obvious move for Jaguar to produce one to captivate an even larger market share.

Producing the estate around the exist-ing X-Type bodyshell meant redesigning everything from the B/C post back; even the "Coke bottle" haunch over the rear wings was removed to ensure a smoothness of line. The "taper" of the saloon's rear wings was also removed so the tailgate opening could be as wide as possible. An increased wraparound of the rear light units helped compensate for this. The roofline was extended and flattened, which meant changes to the door frames. This all resulted in an even stiffer bodyshell.

The tailgate was made up of two sections allowing the whole door to be opened. It was hinged from the top for ease of access, but the upper glass area could be opened separately if needed. Roof rails were fitted to all models finished in either black or silver.

Inside, all the changes focused on the rear compartment/luggage area. New style rear seating had the usual 60/40 split but also could be folded completely down to maximize the loading space. As well as a cargo cover that could be pulled out from a removable cassette by the rear seat, there was an extendable net to affix to points in the roof to protect against luggage moving forward, or as a pet guard. The load area was carpeted with scuff plates and underfloor storage, which included space for a laptop with plug-in point for charging.

Coinciding with the X-Type saloons, minor changes took place like a revised grille and bumper bar, and new steering wheel design.

Mechanically the estate was unaltered from the saloon, except for revised spring rates and other suspension upgrades.

For 2004, history repeated itself with the introduction of a long wheelbase XJ bodyshell for many models. A more integrated design than previous LWB XJs, the new model was only 24kg heavier than the conventional wheelbase car. The extra 5 inches in body length (with a marginal increase in the roof height) all took place after the B/C post, requiring only 17 new body components. The new model wasn't available with the 3.0-liter V-6 engine; otherwise, it was mechanically a conventional XJ apart from some suspension tweaks and a slightly wider turning circle.

Inside, the rear compartment was conventional except for additional legroom. Alternatively, the buyer could opt for a bench seat with four-way electrical lumber adjustment, powered seat recline and headrests, and the ability via a switch to move the front passenger seat forward to provide even more rear legroom. Additional fold-down business

trays could be specified in the rear of the front seats, while screens inset into the rear of the front headrests provided multimedia access.

2005: Another New Diesel

The big news for the X-Type in 2005 was another new engine, a larger 2.2-liter diesel unit, again engineered by Ford, mated to a six-speed manual transmission. Offering 155 brake horsepower with a massive 360 pound-feet of torque at just 1,800 rpm, it provided much-improved performance, with good economy.

Also that year, a Sports Collection Pack became available with a new front spoiler, black grilles, dechroming of exterior trim, revised sill covers, and small boot spoiler, all of which subsequently appeared on a special edition XS model.

Similar treatment was offered for the S-Type models, but this year the 2.5-liter engine was dropped from the range.

A new edition to the XJ range was the Sport Premium, offering a higher degree of specification within the Sport model. Initially available with only the 3.5-liter engine, it later became an option for other models, earlier Sport models eventually being dropped.

2006: XJ Enhancements and a Return of Daimler

At the British Motor Show in 2006, Jaguar introduced their very exclusive XJ Portfolio for the UK market, limited to 100 examples. Effectively a production version of the Concept Eight, and based on the XJR, the car came only in Ultimate Black paint with Ivory leather and alloy interior; it featured highly polished 20-inch Callisto alloy wheels and the wing side power vents in alloy.

Late in the year, anticipating the 2007 MY, various changes took place to the

XJ range. Mesh grilles became a standard fit on all models, windscreen and rear screen moldings were removed with direct sealing of the glass, and the side rubbing strips were discontinued. Heated front seats were standardized on all models and a revised palette of interior and exterior colors were instigated. A tire pressure monitoring system became available as an extra-cost option, along with an automatic speed limiter. Bluetooth connectivity also became available.

There was also a return of the Daimler brand, under pressure from traditional customers. The car was effectively a Super V-8 long wheelbase model only available to special order. With the usual Daimler touches of the fluted grille and badges, and as a top-of-the-range model again, it had all the normal extras as standard equipment including rear business tables and the finest leather and wood trim.

(ABOVE) Jaguar reintroduced the Daimler brand in 2006 for a top-of-the-range X-350, but available only for the UK market and then as a special order.

(RIGHT) The luxurious interior of the Daimler, most of which would later be available for the Jaguar Super V-8, a car that also satisfied this market in the States.

(OPPOSITE) The facelifted XJ with its more aggressive front styling.

2007–2008: The End Is Near

With Jaguar readying to introduce new models to replace their existing saloons, 2007 and 2008 presented the opportunity to make final changes to the range.

All S-Types now gained the front bumper treatment of the R model with the deeper mouth, plus a new range of alloy wheels. The XS that was initially a special edition was now a mainstream model replacing the Sport. Interiors were enhanced with multiadjustable seating with nonperforated leather of a more contemporary style for the SE. Entry models got better trim with Satin Mahogany veneer.

The range was simplified with the Spirit (base model), XS, SE in 2.7-liter diesel or 3.0-liter gasoline, and the 4.2-liter R.

Next, it was time for a facelift of the XJ saloons (now coded X-358). The front end was rejuvenated with a new bumper featuring a deeper pronounced mouth and smaller grilles either side with flush fitting auxiliary lighting. A new radiator grille, again more prominent with a new aggressive growler badge, completed the look.

From the side power vents, like those on the concept car and the XK sports car, appeared in the front wings in body color, while the sills were lowered to match the line of the new bumpers. New door mirrors incorporated indicator lights and new style alloy wheels were available.

At the rear, came another new bumper to match the front, new exhaust tailpipes, a discreet spoiler on the boot lid, and a full-width finisher in chrome or body color.

Internally, there was new seating with softer cushions and, due to the redesign, there was more legroom in the rear compartment. Chrome detailing was applied to a lot of the switchgear and infotainment was enhanced with

The revitalized S-Type nearing the end of production with the XS model.

Bluetooth connectivity. Voice control was now possible with many features.

Finally the XJ range was simplified, based on the XJ Executive in 2.7-liter diesel and 3.0-liter gasoline form (the entry model), XJ Sovereign (2.7 liter/3.0 liter/4.2 liter) high-end model, the XJ Sport Premium (2.7-liter diesel) with sporting touches and stiffer suspension settings, and the XJR (4.2 liter) super-charged model.

Now it was the turn of the X-Type and the biggest change here was the availability of a six-speed automatic transmission with sequential gear change for the 2.2-liter diesel. Although the six-speed manual gearbox was able to

achieve maximum torque in second gear, the new automatic provided that in all gears, ensuring no loss in performance as would normally be the case with automatics.

The revised X-Types now had better communications systems, again with Bluetooth, and improved audio systems.

X-Types received a fresh look, taking cues from the new XJs with the more prominent grille and surround, new growler style badge, new bumper treatment, similar treatment at the rear, and lowered sills, new mirrors and badging.

The interior makeover included a host of new trim options and even the instrument panel got a new look, which

was obviously planned for its replacement model, yet to be seen.

As with the other models, the range was rationalized to 2.0-liter and 2.2-liter diesel car and estate, in X-Type, X-Type S, X-Type Sovereign, and X-Type Sport Premium.

Both the S-Type and X-Type would later give way to Jaguar's new XF range and the XJ would soldier on to 2008 to be deleted a few months prior to the introduction of another new XJ in 2009. A total of 291,000 S-Types, 355,000 X-Types, and 83,500 XJs had been produced.

Last-of-the-line X-Type with various styling changes.

Typical Jaguar XJ interior with dual-colored upholstery, a feature that also became available for the S-Type and X-Type saloons.

SPECIFICATIONS

MODEL	2.5-liter V-6 S-Type	2.7-liter D V-6 S-Type	3.0-liter V-6 S-Type	4.0-liter V-8 S-Type	4.2-liter S-Type	4.2-liter S-Type R
ENGINE SIZE	2,497cc	2,720cc	2,967cc	3,996cc	4,196cc	4,196cc
CARBURETION	Fuel Injection	Fuel Injection	Fuel Injection	Fuel Injection	Fuel Injection	Fuel Injection
MAXIMUM BHP	201@6,800	206@4,000	240@6,800	281@6,100	300@6,000	400@6,100
MAXIMUM TORQUE	184@4,000	320@1,900	221@4,500	287@4,300	310@4,100	408@3,500
GEARBOX	5-speed	n/a	5-speed	n/a	n/a	n/a
AUTOMATIC	5-speed	6-speed	5-speed	5-speed	6-speed	6-speed
0 TO 60 MPH	8.2 sec.	8.1 sec.	6.9 sec.	6.6 sec.	6.2 sec.	5.3 sec.
TOP SPEED	142 mph	143 mph	146 mph	150 mph	155 mph	155 mph
AVERAGE FUEL CONSUMPTION	25.4 mpg	36 mpg	22.7 mpg	24.5 mpg	22.5 mpg	22.5 mpg

MODEL	2.0-liter X-Type	2.0-liter D X-Type	2.2-liter D X-Type	2.5-liter X-Type	3.0-liter X-Type
ENGINE SIZE	2,099cc	1,998cc	2,198cc	2,495cc	2,967cc
CARBURETION	Fuel Injection	Fuel Injection	Fuel Injection	Fuel Injection	Fuel Injection
MAXIMUM BHP	157@6,800	128@3,800	155@3,500	194@6,800	231@6,800
MAXIMUM TORQUE	148@4,100	243@1,800	360@1,800	180@3,000	209@3,000
GEARBOX	5-speed	5 or 6-speed	6-speed	5 or 6-speed	5 or 6-speed
AUTOMATIC	5-speed	n/a	6-speed	5-speed	5-speed
0 TO 60 MPH	8.9 sec.	9.5 sec.	8.5 sec.	7.9 sec.	6.6 sec.
TOP SPEED	130 mph	125 mph	137 mph	140 mph	146 mph
AVERAGE FUEL CONSUMPTION	30.7 mpg	50.3 mpg	47 mpg	29.5 mpg	27.5 mpg

SPECIFICATIONS

MODEL	2.7-liter D X-350	3.0-liter XJ6	3.5-liter XJ6	4.2-liter XJ8	4.2-liter XJ8 R
ENGINE SIZE	2,722cc	2,967cc	3,555cc	4,196cc	4,196
CARBURETION	Fuel Injection	Fuel Injection	Fuel Injection	Fuel Injection	Fuel Injection
MAXIMUM BHP	204@4,000	240@6,800	262@6,250	300@6,000	400@6,100
MAXIMUM TORQUE	321@1,900	221@4,100	254@4,200	310@4,100	408@3,500
GEARBOX	n/a	n/a	n/a	n/a	n/a
AUTOMATIC	6-speed	6-speed	6-speed	6-speed	6-speed
0 TO 60 MPH	7.8 sec.	7.8 sec.	7.3 sec.	6.3 sec.	5 sec.
TOP SPEED	141 mph	145 mph	150 mph	155 mph	155 mph
AVERAGE FUEL CONSUMPTION	35 mpg	27 mpg	26.5 mpg	26 mpg	23 mpg

CHAPTER TWELVE

THE MODERN JAGUARS

Jaguar has been very good at reinventing itself over the years. In the new millennium, the company has modernized its whole outlook, no longer relying on styling generated from the heritage of successful cars gone by.

This process started in 2007 with the introduction of the XF saloon, replacing the X-Type and S-Type models. The XF looked very different from any previous Jaguar saloon and capitalized on the trend for coupe-style four-door saloons, rather than the conventional three-box design. The trend followed with the XJ in 2009.

The F-TYPE sports car, although not so radically different from its predecessor (XK/XKR), took advantage of new technology and smart detailing in design.

In 2013, the XE took Jaguar back into the BMW 3 Series market with a smaller saloon followed by a facelifted XF featuring similar styling.

Jaguar finally entered the SUV market with the F-PACE and then the E-PACE, taking the business into a very lucrative market sector. That was followed by another successful move into the all-electric scene with the I-PACE. Such moves put Jaguar at the forefront of current trends and fashions.

2007–2008: A New Name for a New Car

The XF was announced as a direct replacement for the S-Type and X-Type saloons. Although carrying much of the former car's architecture, including engines and suspension from other current models, body styling was all new—utilizing advanced metals like high carbon and dual-phase hot-formed boron steel.

The striking new front end featured a large mesh grille, vaguely reminiscent of Jaguars from the late 1960s/70s. Distinctive one-piece light clusters, a very shapely bonnet and raked-back windscreen, with a pronounced center roof area to ensure sufficient headroom inside, led to a severe sloping rear window, all giving the appearance of a coupe. The bold styling features of the quite tall rear end emphasized the overall shaping of the body to good effect.

The interior was very modern with alloy paneling, new instrumentation and dashboard layout with some touch-sensitive controls, automatically opening and closing air conditioning vents, and a pronounced center console, the first Jaguar to feature JaguarDrive control with steering wheel mounted

The XF introduced an entirely new styling form for Jaguar, one instigated by their Chief Design Director Ian Callum, which would follow through to all future Jaguar saloons.

gear shift paddles. All other aspects of seating and trim were new to this model.

Initially launched with three engine sizes, 2.7-liter diesel, 3.0-liter gasoline, and 4.2-liter gasoline (plus the supercharged SV-8), trim levels were simplified with Luxury, Premium Luxury (and SV-8).

With an exceptionally good reception to the new car, deliveries took place early on with nearly 2,000 sold in the first few months. Sales continued to be very buoyant into 2008.

2009–2011: New Engines, New Models, a New Car

2009 saw more significant changes to the Jaguar line-up. The XF line got two new engines.

The AJ-V-6D 3.0-liter diesel engine replaced the 2.7-liter unit. Available in two forms, producing 240 or 275 brake horsepower, the new engine offered around 30 percent more power than the old one with 12 percent less emissions and 10 percent improved fuel economy. Its key feature was a parallel sequential

The 3.0-liter V-6 turbo-charged diesel engine developed entirely by Jaguar replaced the joint-venture 2.7-liter diesel.

(ABOVE) The new stylish XJ from 2009 taking the new XF styling features, but to better effect in the longer, lower big car design. The XJ features unique "claw" rear lighting.

(RIGHT) The ultra-modern interior to the XJ with a combination of traditional woods of varying types with other more contemporary finishes. The instrument panel is totally digital, changing display to suit the requirements of particular driving styles.

turbocharging system, the first of its type to be fitted in a V-engine. A new common-rail direct-injection system was also fitted, and by placing the injector crystals deeper in the engine, noise levels were dramatically reduced. Service internals were also extended to 16,000 miles.

This engine brought two new models, the XF diesel and the XF S, the latter with the more powerful version of the engine. The only differentiating points between these and earlier models came with the badging, the standard fit 19-inch alloy wheels, and a boot spoiler.

Jaguar also replaced the 4.2-liter V-8 gasoline engine—with their new 5.0-liter Gen III unit. This was a natural progression for the XF, where it was available in both normally aspirated and supercharged forms (called the XFR). The new 5.0-liter was also available in a new extra-luxury model known as the Portfolio. All post-2009 models received various trim changes, including a larger range of alloy wheels, LED rear lighting, bi-xenon headlights, and a more pronounced frontal panel/radiator area for the XFR.

With the introduction of the revamped XFs, optional Black and Aerodynamic Packs became available to enhance the look of the cars, the latter with reprofiled bumper bars and sills.

Jaguar also launched another entirely new model, a new XJ (coded X-151) in 2009, a direct replacement for the X-350/X-358. This model again had no design and styling carryover from the previous models. The side profile was even more "coupe style" than the XF, but with more flowing lines, and the overall styling suited this larger vehicle.

Unlike the XF, the new XJ was built using Jaguar's latest aluminum technology, making it 150 kilograms lighter than the old (also alloy bodied) XJ. Jaguar also claimed the new car was produced from 50 percent recycled materials and carried a drag coefficient of just

0.29. A unique feature of the new XJ was it full-length panoramic glass roof, helping to create a more streamlined roofline and dramatically enhancing the light entering the cabin. The New XJ was instantly available in both normal and long-wheelbase models. The redesign of the whole bodyshell ensured class-leading standards of ride, handling, and comfort, matched with the use of Jaguar's latest adaptations of suspension and steering, retaining air suspension.

Initially available with a choice of three engines, the 3.0-liter diesel taken from the XF, and the 5.0-liter gasoline in both normally aspirated and super-charged forms. In all cases, it was mated to Jaguar's latest JaguarDrive control with paddle shifts on the steering column, initially with the usual six-speed automatic transmission.

The New XJ was equipped with a vast array of Jaguar's latest technological features, some adapted from other models, like Anti-Lock Braking with Brake Assist, Dynamic Stability Control, Cornering Brake Control, Electronic Traction Control, Engine Drag Torque Control, Pedestrian Contact Sensing System, and the list went on.

Internally, the design was radically different and modern compared to all previous XJs. Controls were of superior quality, many in chromed metal as opposed to plastic. An all-new digital instrument layout replaced conventional dials; the very latest infotainment technology included a touch-sensitive TFT screen with a considerable number of options, including harddrive, plus more modern and extensive trim options.

The cars carried just the XJ insignia but with a choice of trim levels—Luxury, Premium Luxury and Portfolio (3.0-liter diesel), Premium Luxury and Portfolio (5.0 liter), and Portfolio and Supersports (5.0-liter supercharged).

The New XJ was immediately met with great response, proving to be

VCA CERTIFICATION

With the introduction of the X-351, Jaguar received certification from the Vehicle Certification Agency for their comprehensive "cradle to grave" study analyzing the environmental impact of the XJ. This was the first time such a study had been carried out by any manufacturer.

the best performer in terms of overall increased sales in the US against the opposition of BMW, Mercedes, and Lexus. The model went on to receive excellent reports from the famed J.D. Power annual survey.

2012: Model Changes

Various changes and additions took place for the XF in 2012 with the introduction of the 2.2-liter diesel engine in both normal (160 brake horsepower) and S (190 brake horsepower) models. This engine, previously used in the X-Type, was a direct fit into the XF providing a cheaper entry level car with, in some markets, lower taxation. This model introduced an intelligent Stop-Start feature, helping the environment by cutting the engine out when the car was stationary and automatically re-starting the engine when the driver touched the accelerator.

To coincide with this another new model, the SE (Special Equipment), was introduced offering a higher standard of trim along with items like satellite navigation becoming standard.

Another new XF model announced during this period was the Sportbrake,

Jaguar XF Sportbrake, released after the XF saloon to replace the smaller X-Type estate.

an estate car variant to suit the market Jaguar previously held with the X-Type. An excellent five-door model with better luggage capacity than its X-Type equivalent, it was available only in 2.2-liter and 3.0-liter gasoline engine forms, mated to the usual eight-speed transmission. The rear coil-spring suspension from the XF saloons was replaced by an air-suspension system.

The Sportbrake was new from the B/C post back with the extended roofline providing an extra 48 millimeters of headroom for rear passengers, where the seats with a 60/40 split were of a new design. Folding the rear seats down in a one-touch operation afforded a load length of 1,970×1,064 millimeters width. The one-piece tailgate had a one-touch "soft" closure and could also be specified with a powered open/closure opera-

tion. The rear load area had a floor-rail system for the fitment of quick-release storage holdalls and an underfloor area provided extra hidden storage.

The XF Sportbrake proved so successful that Jaguar quoted a 20 percent increase in overall sales, many buyers of which proved to be conquest sales from other brands.

The XJ range came in for revision. New optional Packs could be specified when ordering a new car like the Rear Seat Comfort Pack, providing amenities like electrically adjustable seating and foot rests, on long-wheelbase models. There was also the Sport Pack with aerodynamic trims changes to sills and upgrades to the interior styling. This Pack could also be specified with a Speed Pack (supercharged models only) with a recalibrated

engine-management system, and reworked software.

2013: F for "Fast"

The big news this year was Jaguar's new sports car, the F-TYPE, a proper two-seater sports car and not the grand tourer that was the XK/XKR range. A combination of low weight, excellent aerodynamics, and exemplary performance put Jaguar back in touch with the competition like Porsche's 911.

Following Jaguar's practice now, the bodywork was entirely built of aluminum for lightness and rigidity, with undersill structures and even the boot lid produced from high-strength polymer. Jaguar claimed that 50 percent of the car was produced from recycled materials. Newer AC300 and 6000-series alloys

The F-TYPE sports in all three engine/trim variants.

were specifically chosen for key parts of the car to meet the most stringent requirements for rigidity and lightness. The boot area included a deployable spoiler that automatically raised and folded according to speed to improve aerodynamics and decrease lift.

A new styling approach at the front featured a broad mouth with mesh inset and wide separator bar with the usual growler badge in the center. Vertically formed integrated LED and xenon light units now separated the sports car line from saloons with their horizontally mounted lights. A very aggressive front view was accompanied by a bonnet integral with the front wings and hinged at the front.

From the side, the F-TYPE showed its shorter wheelbase and greater length to the previous models, along with hidden door handles that exposed themselves when required. At the rear, the underbumper area had prominence with different exhaust treatments according to model. Rear light units were horizontal with a hint of a styling nod to the old E-Type.

The interior was also entirely new, yet it retained conventional analog instrumentation as befitting a sports car; all new features and finishes were used.

Two engines could catapult the F-TYPE: the 3.0-liter supercharged V-6 (340 brake horsepower) or, in S form, (380 brake horsepower) the usual 5.0-liter supercharged V-8, all mated to the eight-speed sequential transmission specifically tuned to prioritize acceleration. A mechanical limited-slip differential was fitted to the 3.0-liter model, while the V-8 S got an active

THE SENTINEL

Jaguar introduced their Sentinel long-wheelbase 5.0-liter gasoline model, designed for the executive desiring a high level of in-vehicle security, as an order-only model. Fitted out to individual requirements, the models have been produced using the very latest armor-plating technology, constructed of steels using Kevlar backing, extra-security glass, and specialized equipment levels.

(**ABOVE**) The high-performance XF R with supercharged engine.

(**RIGHT**) The F-TYPE's interior took Jaguar down a different direction to previous sporting models.

electronically controlled differential to maximize traction and provide greater control. Progressively more performance braking systems were fitted according to model, with the ultimate Super High Performance System for the V-8 S including 380-millimeter discs front and 376-millimeter discs at the rear.

For the suspension, an all-alloy double wishbone front and rear design was finely tuned by CAD systems giving lateral stiffness gains of up to 30 percent. A Dynamic Mode enabled the driver at the push of a button to emphasize the sporting characteristics, including throttle response and steering. Jaguar's Adaptive Dynamics was fitted to the S models controlling body roll and pitch rates. Differing exhaust treatments were also fitted according to model.

Initially the F-TYPE was launched just in convertible form, strictly a two-seater with a Z-frame top arrangement that folded down into a neat compartment in the rear quarter without need for a cover. The top could be electrically folded down in just 12 seconds at speeds up to 30 miles per hour. Internal boot space was minimal because of the top and the limited wheelbase, but there was an extra underfloor section for storage.

Another new XF model appeared in 2013. Initially launched at the Los Angeles International Automobile Show the previous November, the XFR-S was said to be the most powerful Jaguar saloon ever produced.

The model embedded the latest high-performance engineering features from the then current XKR-S and new F-TYPE, providing a very driver-focused model. It was fitted with the 550 brake horsepower version of the supercharged 5.0-liter supercharged engine, mated to an eight-seed sequential change automatic transmission, for a 0- to 60-miles-per-hour time of just 4.4 seconds.

Airflow had been carefully managed with some bodywork changes to aid aerodynamics and provide a more aggressive look to the front of the XF, overall lift being reduced by 68 percent. The model also benefitted from extensive suspension and steering modifications.

Over the course of 2012/2013, the range of XF trim options and models expanded to include Sport, LE, and SE Business versions, now totaling 16 models, from the entry-level 2.2-liter diesel SE to the 5.0-liter XFR.

More changes this year for the XJ as an additional model was added, fitted with the F-TYPE's 3.0-liter V-6 gasoline engine developing 340 brake horsepower. This included the new eight-speed sequential change gearbox from the F-TYPE. This eight-speed was also now fitted to other XJ models as well. All XJs now benefited from DAB radio systems and an improved Meridian hi-fi system.

In the same year, for specific overseas markets like China, the States, Russia, and continental Europe, a 3.0-liter 340-brake-horsepower XJ and XF became available with all-wheel-drive. The USA was the world's leading market for all-wheel-drive sales and three quarters of all AWD XJs ended up in the States.

At the New York Motor Show that year, a new top-of-the-range XJ was announced, reintroducing the XJR insignia. Using the 5.0-liter supercharged engine, this car now incorporated many of the suspension, steering, and brake systems from the F-TYPE sports car, along with interior trim changes to differentiate this from other XJ models.

Another new model was the Ultimate, fitted with every conceivable extra to a long-wheelbase model in both 3.0-liter diesel and 5.0-liter gasoline forms. By the end of 2013, the 3.0-liter diesel XJ short wheelbase models were discontinued.

AWARDS KEEP COMING

Even since the launch of the XF in 2007, the model had been receiving awards from around the world. In 2013, it was the Car of the Decade Award made by the Southern Group of Motoring Writers (SGMW), decided on by a team of 30 renowned motoring journalists, editors, and authors.

Later the XF was voted The Business Car's Executive Car of the Year and the Luxury Car of the Year. In total, the model received over 100 coveted awards over the years.

2014: Tin-Top F-TYPE

An XFR-S Sportbrake joined the range of models, incorporating all the features of the saloon version, plus a neat spoiler over the rear hatch.

Another new optional Package became available, the R-Sport Pack, just for the 2.2-liter models, with a range of trim changes and enhancements to ride settings. Generally trim and equipment levels were improved on all cars to match similar changes to other models, but the overall range was reduced in readiness for a fall-off in production to make way for a total facelift to the XF. The SE, SE Business, and Premium Luxury models were dropped as were both the 3.0-liter and 3.0-liter diesel engine options. The Sportbrake models in both 2.2-liter diesel and 5.0-liter gasoline forms continued into 2015.

The XJ R-Sport model was introduced following styling and trim

PROJECT 7 CONCEPT

At the British Goodwood Festival of Speed in 2013, Jaguar announced their Project 7 concept, based on the F-TYPE—the name acknowledging Jaguar's seven victories at the Le Mans 24-hour race.

External design changes included a fairing behind the driver's head, bespoke carbon-fiber components, a new front splitter in the grille, side skirts, and a real diffuser. The windscreen was lowered and there was a new front bumper arrangement.

Internally, a composite single seat had a racing harness, a helmet holder, and custom trim.

The project was a fully functioning prototype fitted with the 550-brake-horsepower 5.0-liter supercharged V-8 engine and other drivetrain aspects from the F-TYPE.

The car finally made it to production as the Project 8, a two-seater road car with a limited run of just 250, hand built by Jaguar Land Rover Special Operations Team. Each car carried a numbered plaque fitted between the seats signed by Jaguar's Chief Designer Ian Callum.

The performance of the 5.0-liter engine was enhanced to provide 575 brake horsepower mated to the eight-speed quickshift transmission, presenting a 0- to 60-miles-per-hour acceleration time of 3.9 seconds.

Bodily, the "production" car was a mirror of the concept but with two-seats with rollover hoops and a removable roof.

adaptations taken from other Jaguar saloons of the period.

A coupe version of the F-TYPE became available in February—for many, a much more stylish model than the convertible. With its distinctive sweeping roofline leading to a single tailgate hinged at the top, with this model it was possible to order a full-length panoramic glass roof. Jaguar claimed this model was the most torsionally strong bodyshell they had ever produced. Rear styling was exceptionally clean and attractive, unmistakably Jaguar and again with influences from the original E-Type. Internally, luggage space was dramatically improved with 407 liters of usable space.

By November of this year, the F-TYPE range had been expanded significantly to include 14 models and the renaming of one, the V-8 S becoming the F-TYPE R! All models were available in both coupe and convertible forms except for the new F-TYPE R (coupe only). This car now incorporated the 550-brake-horsepower version of the V-8 engine with appropriate changing to all the car's dynamics to match the increased performance. The model was also available with carbon ceramic matrix brakes.

2015: Return of the Compact Jaguar

The XE saloon was a very important new model announced by Jaguar with deliveries starting this year. Effectively a replacement for the X-Type and to supplement the XF, which was too big a car for many in this market, the XE provided Jaguar's best competitive element against BMW's 3 Series.

Built extensively from aluminum, some parts of which were produced from RC5754-grade, a high-strength alloy used for the first time in a car, the model also featured riveting and bonding methods devised for the XJ. The XE formed part

(ABOVE) Released after the F-TYPE roadster, the coupe is a very attractive variant with more practical load capacity and a smart aerodynamic style, from the rear showing remembrances of the 1960s E-Type.

(LEFT) One of the luxuriously appointed XJ models, now with quilted upholstery and many other features to maintain the car's credibility as a top-of-the-range luxury limousine.

Jaguar XE model, the final replacement for the X-Type incorporating the current styling trends of other Jaguar saloons but in a smaller, more economical, and efficient package.

One of the new highly efficient Ingenium engines produced in-house at Jaguar Land Rover's new facility in Wolverhampton in the West Midlands of the UK.

The system also incorporated Torque Vectoring, first seen on the F-TYPE sports car.

The XE was powered by a range of aluminum four- and six-cylinder gasoline and diesel engines from Jaguar's entirely new Ingenium range, built at their new facility in Wolverhampton in the West Midlands. All these engines benefitted from direct fuel injection, variable valve timing, and intelligent stop/start system. Engine options were a 2.0-liter turbocharged diesel (163 and 180 brake horsepower), 2.0-liter turbocharged gasoline (200 and 240 brake horsepower), plus the 3.0-liter supercharged gasoline engine derived from the F-TYPE (340 brake horsepower), although the more powerful engines didn't come to market until later. All models were offered in a choice of trim levels, SE (Special Equipment), Prestige, R-Sport, or Portfolio, with just an S model reserved for the 3.0-liter engine car.

Engines were mated to a conventional six-speed manual with a lightweight aluminum sump and semidry sump lubrication for all 2.0-liter engines and the ZF 8HP eight-speed automatic transmission for gasoline engine cars.

Internally, the XE was another all-new design with the latest Infotainment system and modern approaches to a wide range of wood and alloy veneers and leather upholstery, all styles that would follow through to other Jaguar models in due course. Depending on model and specification, all the latest technologies devised for other Jaguars were incorporated in the XE, the first Jaguar to be built at the Rover Solihull factory (now also part of the Tata organization) in the West Midlands.

In the same year, the XF came in for a total facelift with bodywork being designed and produced around an aluminum-intensive structure and with styling remarkably similar to its smaller counterpart. This made the "new" XF

of a modular construction design that would ultimately spawn extra models.

Derived from the XF design with a lower profile with a lightweight underfloor panel to reduce drag, the XE recorded the lowest drag co-efficient of any Jaguar at 0.26.

The suspension package was based on double front wishbones and an integrated link at the rear, keeping weight to a minimum with extensive aluminum usage in the toe links and upper control arms. The XE was the first Jaguar to use electric power steering. The braking system had large discs (ventilated at the front) with suspension-mounted ducts to aid brake cooling and a new electronic brake-system controller.

(**ABOVE**) A completely new interior for the XE, a general style that would follow through to the later XF and SUV models.

(**LEFT**) The new XF, more than a facelift of the old model, now made from aluminum and with resemblances to its smaller sibling, the XE.

80kg lighter than the old model and 28 percent stiffer. The car encompassed the latest developments in crash protection and passenger safety. The new styling brought an increase of 51 millimeters in the car's wheelbase, while the body got 7 millimeters shorter and 3 millimeters lower in height.

The interior was also new and again took a lot from the XE in terms of overall design, equipment and specification. Not only had the new XF the looks of the XE and the interior, but also most of the technical advances from that model and other Jaguar cars.

The new XF was initially available with all the engine choices from the XE plus the 3.0-liter V-6 diesel and gasoline engines. Trim levels were also the same as the XE, making a total of 14 models in the new XF range.

More changes took place for the F-TYPE range. All-wheel-drive became an option for all models except the 3.0-liter 340 brake horsepower car. A six-speed fully manual ZF gearbox with dual mass flywheel was now an extra choice on rear-wheel-drive models, and the F-TYPE R was now available in convertible form.

Certain changes took place to trim and interior levels including Jaguar's latest InControl and Infotainment systems to match other models.

2016: SUV Time

For 2016, the whole XJ was reconfigured, taking advantage of technological advances and comfort levels. As well as the existing models, two new variants became available, the 3.0-liter diesel and 5.0-liter gasoline Autobiography.

At the front, a new larger, more upright and deeper grille was used, air intakes had redesigned chromed blades and "J-blade" LED daytime lighting was fitted to new lighting units. "J-signature" rear lighting also appeared,

the rear bumper shaping was altered, there was new badging, and new shaping to the exhaust tailpipes.

Inside all specifications were upgraded with some items previously available only on Portfolio, including a new style quilted-leather trim for some cars. The Autobiography (long-wheelbase only) had the ultimate equipment level with features like built-in rear TV/DVD/computer screens, rear multimedia system, and individualized electrically reclining seats. With these changes the old Supersports and Ultimate models were discontinued.

F-TYPE name changes were made. The V-6, and V-6 S were replaced by just F-TYPE or F-TYPE S. An electrically boosted rack-and-pinion steering system was fitted and yet more trim upgrades.

More significant however was the introduction of the F-TYPE SVR, an all-weather supercar for everyday use, with a top speed over 200 mph and a 0- to 60-miles-per-hour time of just 3.5 seconds for a price of £115,485 (convertible) or £110,000 (coupe). The car featured the latest evolution of the 5.0-liter supercharged engine with significant changes in calibration achieving 575 brake horsepower, with powertrain efficiency increased through enlarged air intakes in the front bumper, revised air coolers, and redesigned bonnet vents. A lightweight titanium exhaust system decreased overall weight and reduced back pressure.

To accompany the engine changes the eight-speed quick-change gearbox was recalibrated, and better traction was assured by tires 10mm wider than those fitted to the F-TYPE R, on weight-saving alloy wheels. Optimum torque distribution to both axles was assured by refining the electronic active differential. Other changes included a thicker rear anti-roll bar, revised damper control, and a new rear knuckle to the suspension.

The SVR was 25 kilograms lighter than the standard all-wheel-drive F-TYPE R, or 50 kilograms lighter if fitted with the carbon-ceramic matrix braking system. To match all the external and drivetrain changes, the interior received a unique finish for this model.

XEs got their first revisions in 2016 for the 2017 model year, to feature many of the changes affecting other models, such as Infotainment.

The hot news for 2016 was the introduction of not just a new Jaguar model, but an entirely new area of the market that the company had never explored before, but which was becoming ever more popular worldwide, the SUV (Sports Utility Vehicle). A two-way communications channel between Jaguar and Land Rover (both companies now owned by the Indian Tata group) obviously aided the development of this concept.

Called the F-PACE, it was Jaguar's first attempt at a performance crossover vehicle, already the first in its class to be constructed 80 percent in aluminum, with additional weight savings from a composite tailgate and magnesium crossover beam. Other firsts included boot volume (650 liters), cabin width, near compartment knee-room, and ease of access. The F-PACE shared no other chassis dimensions with other Jaguars; it did feature stunning good looks encompassing Jaguar's now "corporate" style frontage and clean lines and roof. The car was equipped with all-wheel-drive and most of the technically advanced features Jaguar offered for other models. A new feature was the Activity Key, a waterproof wearable technology allowing for the keys to the car to be securely locked in the vehicle.

The powertrain for the F-PACE included the Ingenium 2.0-liter 180-brake-horsepower engine with both rear and all-wheel-drive with manual or automatic transmission, and the 3.0-liter supercharged V-6 gasoline and 3.0-liter

(ABOVE) The ultra-high-performance F-TYPE SVR.

(LEFT) Jaguar's first venture into the SUV market with the F-PACE.

The quickest of all the XEs, the Special Vehicle Operations XE SVO.

diesel engines with automatic transmission and all-wheel-drive. Cars were available in Prestige, R-Sport, Portfolio, and S trim level, and, initially, 2,000 First Edition examples exceptionally well specified, with either of the 3.0-liter engines.

2017: Aggressive Upgrades

In June, Jaguar started production of the more powerful Ingenium engines, the 240-brake-horsepower 2.0-liter four-cylinder twin-turbo diesel, a strengthened engine with uprated pistons, crankshaft, and fuel injection. A first was the exhaust manifold integrated with the cylinder head casting, passing coolant through the manifold, reducing warm-up time. This engine was available for the XE, XF, and now a new 2.0-liter F-TYPE adding two additional models to the sports car range.

All models came in for revision, starting with the XE. A high-powered

SV Project 8 was announced following the equivalent F-TYPE mentioned earlier. Also, utilizing carbon fiber for lightness and extensive modifications to the drivertrain, this became a limited-edition (300 cars) 200-miles-per-hour sporting saloon.

For those looking for more performance from the conventional models, the XE S was announced, sharing the F-TYPE's 3.0-liter 380-brake-horsepower gasoline engine with configurable driver dynamics to enhance the driving pleasure. Other XEs benefitted from equipment upgrades as well.

Along with the more powerful engines fitted to the XF, configurable dynamics was standardized with better equipment levels. A Sportbrake version finally became available from the new bodyshell.

A whole range of upgrades also appeared for the flagship XJ range, involving new trim levels and more

advanced technical features both for the drivetrain and infotainment systems. All models were now fitted with the uprated more compact ZF eight-speed 8HP45 transmission.

Now a new model appeared, the XJR575, using a 575-brake-horsepower 5.0-liter supercharged engine delivering a 0- to 60-miles-per-hour time of a mere 4.2 seconds. Available only on the short-wheelbase body, it was offered in two distinct colors, Velocity Blue or Satin Corris Grey. It also featured a new rear spoiler, side sills, new front bumper and lower air intakes, bonnet vents, and 20-inch Farallon alloy wheels. Internally, diamond-stitched leather seating and "575" branding differentiated this from other XJs.

For the F-TYPE sports car, Jaguar broadened its appeal, firstly by fitting the 300-brake-horsepower 2.0-liter Ingenium engine as an additional model, still offering excellent performance (0 to

(ABOVE) The best of the XJs, the XJ575 indicating its 575-brake-horsepower performance engine.

(LEFT) The unique and racey interior to Jaguar's fastest four-door model, the XE SVO.

60 miles per hour in 5.4 seconds), with unique styling features at a low price.

Even the F-PACE came in for revision, now including the 163-brake-horsepower four-cylinder engine with manual transmission and rear-wheel drive. In common with the changes to other Jaguar models, the F-PACE also received upgraded trim and systems.

2018: Crossing Over

Another major new model from Jaguar was the E-PACE, a five-seater compact SUV to complement the larger F-PACE.

Bodily the E-PACE was not all aluminum, retaining steel for bodysides, though of just 0.7 millimeter thickness for lightness. The bonnet, front wings, roof panel, and tailgate were all aluminum, while the crossbeam followed Jaguar's current practice of using magnesium.

Although some aspects of the underside were decidedly based on the Land Rover Evoque, the front suspension included a new subframe to generate greater stiffness with solid rear mounts to contribute a more Jaguar feel to the steering and handling. The rear suspension was very compact and based around that developed for the F-PACE. All the now usual features including adaptive dynamics, active driveline all-wheel-drive, and interior features and equipment levels were present.

The E-PACE was powered by a choice of the Ingenium four-cylinder diesel engines of 150 brake horsepower, 180 brake horsepower, or 240 brake horsepower, or the 249-/300-brake-horsepower gasoline engines, all mated to a new ZF nine-speed automatic, according to trim with steering-wheel mounted paddles. For the less powerful engines, there was also the option of a six-speed manual gearbox.

E-PACE models were launched in S, SE, and HSE trim levels with the choice of the five powertrains. A Dynamic R version was also offered with external trim design upgrades, sports seating, and enhanced detailing. There was also a First Edition model, offered only during the first year of production, again with additional trim finishes and added features from the Dynamic R and SE models.

Jaguar's attack on new markets with SUVs was successful, but even more important, looking to the future and environmental issues, the company launched in 2018 the I-PACE, the first all-electric vehicle, a direct competitor to the likes of the Tesla X.

A sleek crossover between an SUV and sporting vehicle, it has around the same wheelbase as an XE saloon, but with a lot more space inside without the need for traditional mechanical packaging. There's 890 millimeters of legroom in the rear, and 10.5 liters of storage because of a lack of a transmission tunnel, plus 656 liters of rear luggage accommodation or 1,453 liters with the rear seats folded.

Featuring a traditional Jaguar front with a short bonnet, aero-enhanced roof design, and curved rear screen, the cab-forward styling contrasts with the squared off rear reducing drag. Active vanes in the front grille allow cooling airflow when required for the batteries and climate control. The car is built using Jaguar's now well-known aluminum skills including riveting and bonding.

The I-PACE uses a state-of-the-art 90-kilowatt lithium-ion battery using 432 pouch cells, which could deliver up to 298 miles of charged range. A 0 to 80 percent charge can be carried out in 85 minutes using 50-kilowatt DC charging and is fully compatible with DC Rapid Chargers (100 kilowatt). Home charging via an AC wall box (7 kilowatt) can be carried out in around 10 hours.

The car's drive system is based on two electric motors featuring driveshafts passing through them and placed near each axle. They produce a combined performance of the equivalent of 400 brake horsepower and 696N-m of torque, via the all-wheel-drive system. In performance terms, the I-PACE can launch from 0 to 60 miles per hour in just 4.5 seconds.

The I-PACE is a tour de force in electronics, the first Jaguar to feature a digital "flight deck" infotainment system incorporating two touchscreens, designed to minimize distraction for the driver, performing all instructions logically. A heads-up display keeps the driver informed of key elements and there's also an advanced satellite navigational system. A range of "smart settings" automatically adjust the car's controls to suit the driver, who via the In-Control apps can access Amazon's Alexa. Software updates for the car's systems can be received directly via wi-fi.

Four models are currently available, the S, SE, HSE, and a First Edition, following similar trim and features to Jaguar's SUV models.

Jaguar's E-PACE and I-PACE vehicles are not produced in any of the group's UK-based factories. Instead the E-PACE is assembled at the Magna Steyr factory in Graz, Austria, but also, exclusively for the Chinese market, at Chery Jaguar Land Rover's state-of-the-art facility in Changshu.

This brings the Jaguar story up to date. A British company that has sustained many issues and owners over nearly a century of production, commencing with simple sidecars, has evolved to produce some of the highest performing and most technologically advanced vehicles for a global market.

(TOP) Jaguar's launch into the all-electric market, the I-PACE. **(BOTTOM, LEFT)** The stylish and very modern approach to the I-PACE interior. **(BOTTOM, RIGHT)** Jaguar's smaller SUV/crossover, the E-PACE.

SPECIFICATIONS

XF (TO 2016)	2.2-liter D XF	2.2-liter D XF S	27-liter D XF	3.0-liter P XF	3.0-liter P XF (2013 on)	3.0-liter D XF
ENGINE SIZE	2,179cc	2,179cc	2,720cc	2,967cc	2,993cc	2,993cc
CARBURETION	Fuel Injection	Fuel Injection	Fuel Injection	Fuel Injection	Fuel Injection	Fuel Injection
MAXIMUM BHP	163@3,500	190@3,500	207@4,000	238@6,800	340@6,000	240@4,000
MAXIMUM TORQUE	295@2,000	332@2,000	320@1,900	216@4,100	332@5,000	369@2,000
AUTOMATIC	8-speed	8-speed	6-speed	6-speed	8-speed	6- or 8-speed
0 TO 60 MPH	9.8 sec.	8 sec.	7.7 sec.	7.9 sec.	6 sec.	6.7 sec.
TOP SPEED	130 mph	140 mph	143 mph	148 mph	155 mph	149 mph
AVERAGE FUEL CONSUMPTION	52.3 mpg	52 mpg	37.6 mpg	26.8 mpg	29 mpg	42 mpg

XF (TO 2016) continued	3.0-liter D XF S	4.2-liter P XF	4.2-liter P XF R	5.0-liter P XF	5.0-liter P XF R	5.0-liter P XFR-S
ENGINE SIZE	2,993cc	4,196cc	4,196cc	5,000cc	5,000cc	5,000cc
CARBURETION	Fuel Injection	Fuel Injection	Fuel Injection	Fuel Injection	Fuel Injection	Fuel Injection
MAXIMUM BHP	275@4,000	298@6,000	416@6,250	385@6,500	510@6,500	550@6,500
MAXIMUM TORQUE	443@2,000	303@4,100	413@4,000	380@3,500	461@5,500	502@5,500
AUTOMATIC	6-speed	6-speed	6-speed	6- or 8-speed	6- or 8-speed	8-speed
0 TO 60 MPH	5.3 sec.	6.2 sec.	5.1 sec.	5.5 sec.	4.7 sec.	4.4 sec.
TOP SPEED	155 mph	155 mph	155 mph	155 mph	155 mph	186 mph
AVERAGE FUEL CONSUMPTION	22.5 mpg	25 mpg	22 mpg	25 mpg	24 mpg	24.2 mpg

NEW XF	2.0-liter D RWD	2.0-liter D RWD	2.0-liter D AWD	2.0-liter D RWD	2.0-liter D AWD	3.0-liter V-6 D RWD
ENGINE SIZE	1,999cc	1,999cc	1,999cc	1,999cc	1,999cc	2,993cc
CARBURETION	Fuel Injection	Fuel Injection	Fuel Injection	Fuel Injection	Fuel Injection	Fuel Injection
MAXIMUM BHP	163@4,000	180@4,000	180@4,000	240@4,000	240@4,000	300@4,000
MAXIMUM TORQUE	280@3,500	317@2,500	317@2,500	500@1,500	500@1,500	700@2,000
GEARBOX	6-speed	6-speed	6-speed	6-speed	6-speed	n/a
AUTOMATIC	8-speed	8-speed	8-speed	8-speed	8-speed	8-speed
0 TO 60 MPH	8.2 sec.	7.5 sec.	7.7 sec.	6.3 sec.	6.4 sec.	6.1 sec.
TOP SPEED	132 mph	136 mph	136 mph	153 mph	150 mph	155 mph
AVERAGE FUEL CONSUMPTION	68.9 mpg	65.7 mpg	65.7 mpg	53.3 mpg	48.7 mpg	49.6 mpg

SPECIFICATIONS

NEW XF *continued*	2.0-liter P RWD	2.0-liter P RWD	2.0-liter P AWD	2.0-liter P AWD	3.0-liter P RWD
ENGINE SIZE	1,997cc	1,997cc	1,997cc	1,997cc	2,995cc
CARBURETION	Fuel Injection	Fuel Injection	Fuel Injection	Fuel Injection	Fuel Injection
MAXIMUM BHP	200@5,500	250@5,500	250@5,500	300@5,500	380@6,500
MAXIMUM TORQUE	320@1,200	365@4,000	365@4,000	400@4,500	450@4,500
GEARBOX	6-speed	6-speed	6-speed	n/a	n/a
AUTOMATIC	8-speed	8-speed	8-speed	8-speed	8-speed
0 TO 60 MPH	7.1 sec.	6.1 sec.	6.2 sec.	5.5 sec.	5.1 sec.
TOP SPEED	146 mph	152 mph	147 mph	155 mph	155 mph
AVERAGE FUEL CONSUMPTION	41.5 mpg	41.5 mpg	40.9 mpg	40 mpg	34.4 mpg

XJ (X-351)	2.2-liter D	3.0-liter P	3.0-liter D	3.0-liter D	5.0-liter P
ENGINE SIZE	2,219cc	2,995cc	2,993cc	2,993cc	5,000cc
CARBURETION	Fuel Injection	Fuel Injection	Fuel Injection	Fuel Injection	Fuel Injection
MAXIMUM BHP	190@3,500	335@6,500	275@4,000	296@4,000	385@6,500
MAXIMUM TORQUE	332@2,000	332@5,000	400@2,000	516@2,000	380@3,500
AUTOMATIC	8-speed	8-speed	6- or 8-speed	8-speed	6- or 8-speed
0 TO 60 MPH	n/a	5.7 sec.	6 sec.	5.9 sec.	5.4 sec.
TOP SPEED	n/a	155 mph	155 mph	155 mph	155 mph
AVERAGE FUEL CONSUMPTION	n/a	31 mpg	49 mpg	49 mpg	24 mpg

XJ (X-351) *continued*	5.0-liter P Supersports	5.0-liter P XJR	5.0-liter P XJR575
ENGINE SIZE	5,000cc	5,000cc	5,000cc
CARBURETION	Fuel Injection	Fuel Injection	Fuel Injection
MAXIMUM BHP	503@6,500	543@6,500	575@6,500
MAXIMUM TORQUE	461@5,500	502@5,500	700@3,520
AUTOMATIC	6-speed	8-speed	8-speed
0 TO 60 MPH	4.7 sec.	4.4 sec.	4.2 sec.
TOP SPEED	155 mph	174 mph	186 mph
AVERAGE FUEL CONSUMPTION	25 mpg	25 mpg	24 mpg

SPECIFICATIONS

F-TYPE	2.0-liter D RWD	2.0-liter D RWD	2.0-liter D AWD	2.0-liter D AWD	3.0-liter D V-6 RWD	2.0-liter P V-4 RWD
ENGINE SIZE	1,999cc	1,999cc	1,999cc	1,999cc	2,993cc	1,997cc
CARBURETION	Fuel Injection	Fuel Injection	Fuel Injection	Fuel Injection	Fuel Injection	Fuel Injection
MAXIMUM BHP	163@4,000	180@4,000	180@4,000	240@4,000	300@4,000	200@5,500
MAXIMUM TORQUE	380@2,500	430@2,500	430@2,500	500@1,500	700@2,000	320@4,500
GEARBOX	6-speed	6-speed	6-speed	6-speed	n/a	n/a
AUTOMATIC	8-speed	8-speed	8-speed	8-speed	8-speed	9-speed
0 TO 60 MPH	8.2 sec.	7.5 sec.	7.7 sec.	6.3 sec.	5.8 sec.	7.1 sec.
TOP SPEED	132 mph	136 mph	136 mph	153 mph	155 mph	146 mph
AVERAGE FUEL CONSUMPTION	68.9 mpg	65.7 mpg	65.7 mpg	53.3 mpg	51.4 mpg	41.5 mpg

F-TYPE continued	2.0-liter P V-4 RWD	2.0-liter P V-4 RWD	3.0-liter P V-6 RWD
ENGINE SIZE	1,997cc	1,997cc	2,995cc
CARBURETION	Fuel Injection	Fuel Injection	Fuel Injection
MAXIMUM BHP	250@5,500	300@5,500	380@6,500
MAXIMUM TORQUE	365@4,500	400@4,500	450@4,500
GEARBOX	n/a	n/a	n/a
AUTOMATIC	9-speed	9-speed	9-speed
0 TO 60 MPH	6.3 sec.	5.5 sec.	5.1 sec.
TOP SPEED	152 mph	155 mph	155 mph
AVERAGE FUEL CONSUMPTION	41.5 mpg	40 mpg	34.4 mpg

XE	2.0-liter D	2.0-liter D AWD	2.0-liter P	2.0-liter P	3.0-liter P
ENGINE SIZE	1,999cc	1,999cc	1,999cc	1,999cc	2,995cc
CARBURETION	Fuel Injection	Fuel injection	Fuel Injection	Fuel Injection	Fuel Injection
MAXIMUM BHP	163@4,000	180@4,000	200@5,500	240@5,500	340@6,500
MAXIMUM TORQUE	280@2,500	317@2,500	206@4,000	250@4,000	312@4,500
GEARBOX	6-speed	6-speed	6-speed	6-speed	n/a
AUTOMATIC	8-speed	8-speed	8-speed	8-speed	8-speed
0 TO 60 MPH	7.9 sec.	7.9 sec.	7.1 sec.	6.5 sec.	4.9 sec.
TOP SPEED	132 mph	140 mph	147 mph	155 mph	155 mph
AVERAGE FUEL CONSUMPTION	75 mpg	67 mpg	38 mpg	38 mpg	34.9 mpg

SPECIFICATIONS

F-PACE	2.0-liter D RWD	2.0-liter D RWD	2.0-liter D AWD	2.0-liter D AWD
ENGINE SIZE	1,999cc	1,999cc	1,999cc	1,999cc
CARBURETION	Fuel Injection	Fuel Injection	Fuel Injection	Fuel Injection
MAXIMUM BHP	163@4,000	180@4,000	180@4,000	240@4,000
MAXIMUM TORQUE	380@1,750	317@2,500	317@2,500	500@1,500
GEARBOX	6-speed	6-speed	6-speed	6-speed
AUTOMATIC	8-speed	8-speed	8-speed	8-speed
0 TO 60 MPH	9.7 sec.	8 sec.	8.2 sec.	6.1 sec.
TOP SPEED	121 mph	129 mph	129 mph	135 mph
AVERAGE FUEL CONSUMPTION	57.7 mpg	55.4 mpg	54.3 mpg	48.7 mpg

F-PACE continued	2.0-liter P AWD	3.0-liter V-6 D AWD	3.0-liter V-6 P AWD
ENGINE SIZE	1,997cc	2,993cc	2,995cc
CARBURETION	Fuel Injection	Fuel Injection	Fuel Injection
MAXIMUM BHP	250@5,500	300@4,000	380@6,500
MAXIMUM TORQUE	365@2,500	700@1,750	450@4,500
GEARBOX	6-speed	n/a	n/a
AUTOMATIC	8-speed	8-speed	8-speed
0 TO 60 MPH	6.4 sec.	5.8 sec.	5.1 sec.
TOP SPEED	135 mph	150 mph	155 mph
AVERAGE FUEL CONSUMPTION	38.2 mpg	47.1 mpg	31.7 mpg

SPECIFICATIONS

E-PACE	2.0-liter D FWD	2.0-liter D FWD	2.0-liter D FWD	2.0-liter D AWD
ENGINE SIZE	1,999cc	1,999cc	1,999cc	1,999cc
CARBURETION	Fuel Injection	Fuel Injection	Fuel Injection	Fuel Injection
MAXIMUM BHP	150@3,500	163@4,000	180@4,000	180@4,000
MAXIMUM TORQUE	180@1,750	380@2,500	430@1,750	430@1,750
GEARBOX	6-speed	6-speed	6-speed	n/a
AUTOMATIC	9-speed	9-speed	9-speed	9-speed
0 TO 60 MPH	9.5 sec.	9.7 sec.	8 sec.	8.2 sec.
TOP SPEED	124 mph	121 mph	129 mph	129 mph
AVERAGE FUEL CONSUMPTION	55.7 mpg	57.7 mpg	55.4 mpg	54.3 mpg

E-PACE *continued*	2.0-liter D AWD	2.0-liter P AWD	3.0-liter D AWD	3.0-liter P AWD
ENGINE SIZE	1,999cc	1,997cc	2,993cc	2,995cc
CARBURETION	Fuel Injection	Fuel Injection	Fuel Injection	Fuel Injection
MAXIMUM BHP	240@4,000	250@5,500	300@4,000	380@6,500
MAXIMUM TORQUE	500@1,500	365@4,000	700@1,750	450@4,500
GEARBOX	n/a	n/a	n/a	n/a
AUTOMATIC	9-speed	9-speed	9-speed	9-speed
0 TO 60 MPH	6.7 sec.	6.4 sec.	5.8 sec.	5.1 sec.
TOP SPEED	135 mph	135 mph	150 mph	155 mph
AVERAGE FUEL CONSUMPTION	48.7 mpg	38.2 mpg	47.1 mpg	31.7 mpg

INDEX

IMAGE CREDITS

b=bottom, l=left, m=middle, r=right, t=top

Author's Images (images directly attributable to images taken by the author or acquired): 5, 11t, 13t, 14t, 14b, 15, 18tl, 18tr, 19m, 20m, 21tl, 21tr, 24, 25tl, 26m, 27, 28t, 28b, 29, 30, 32tl, 32tr, 32m, 32b, 33t, 34-35, 36tl, 36tr, 36b, 37t, 37b, 38, 39, 40t, 40m, 40b, 43m, 44b, 46tl, 48t, 48b, 49t, 49b, 51tl, 51tr, 51m, 52, 54tl, 54m, 54b, 55, 56t, 57t, 58, 59, 61t, 61b, 64b, 70l, 70r, 72t, 73b, 74, 77ml, 77mr, 77b, 80tl, 80br, 81m, 83b, 84t, 84b, 85t, 85b, 86t, 86m, 88l, 88r, 90t, 90bl, 90br, 91, 98b, 99t, 99m, 99b, 100, 101t, 102t, 102b, 104b, 106t, 107, 109t, 109m, 110t, 110b, 111, 112t, 112b, 113t, 114, 118t, 121, 126t, 126b, 127b, 128t, 128b, 129t, 129b, 131t, 131b, 133, 134t, 134b, 125t, 125m, 135t, 135b, 136, 139t, 139b, 140t, 140b, 141t, 141m, 141b, 144t, 144b, 147t, 147b, 150, 151t, 151b, 152t, 152b, 154, 155r, 157b, 158b, 164r, 166b, 174b, 176tr, 178t, 180t, 180b, 181b, 182, 187, 189b, 190, 192, 194t, 194b, 197t, 199, 201t, 214b, 216, 218t, 219, 221t, 234, 240t, 240b, 243b, 247br. **Author's Collection** (images within the author's own archive collection, including copies supplied for his use, and those he has copied from period brochures/documentation that he owns): 9, 11b, 12, 13b, 18m, 19t, 20t, 23t, 23b, 26t, 34, 33m, 46bl, 53b, 57b, 63b, 62, 67b, 69t, 67t, 72m, 72b, 73t, 77t, 89, 82, 83t, 87t, 95, 96, 101b, 105tl, 105tr, 108tr, 108m, 113m, 142, 146, 149t, 149b, 153, 156, 158t, 163t, 166t, 168b, 169, 172, 173l, 173r, 176b, 195, 198bl, 198br. **Jaguar Cars/Author's Collection** (images previously supplied to the author for press use by Jaguar Cars/Jaguar Land Rover): 6-7, 56m, 120, 127t, 130, 143, 145tl, 145tr, 148, 155tl, 157t, 161, 162t, 162b, 164l, 165, 168t, 171, 174t, 175t, 175b, 176tl, 178m, 179, 181t, 183, 188, 189t, 191t, 191b, 197b, 198t, 201bl, 201br, 202, 203, 204, 205t, 205b, 206-207, 208, 211, 212t, 212b, 214t, 215, 218b, 220, 222, 224t, 224b, 225, 226, 227t, 227b, 231t, 231b, 232t, 232b, 235, 236t, 236b, 238, 239t, 239b, 241t, 241b, 243t, 244, 245t, 245b, 247t, 247bl, back cover. **Jaguar Heritage/Author's Collection** (images previously supplied to the author from Heritage): 17t, 17b, 25tr, 25b, 31, 43b, 44t, 45, 46tr, 46br, 53t, 54tr, 64t, 69b, 75, 78, 76t, 76b, 80bl, 87b, 81t, 97, 98t, 104t, 106b, 108tl, 118b, 119b, 124, 123t, 123b, 163b. Jaguar Heritage/Paul Skilleter Collection: 117. James Mann: dustjacket front, front and back cover boards, 2-3. Graham Searle: 50. Shutterstock: front endpapers (Radu Bercan), back endpapers (Lerner Vadim). Unknown Source (within author's collection): 68, 79b, 119t